BY APPOINTMENT
TO H.M. QUEEN ELIZABETH II
BRITISH LEYLAND UK LIMITED
LEYLAND CARS
MANUFACTURES OF ROVER CARS
LAND ROVERS, RANGE ROVERS
AND AUSTIN CARS
BIRMINGHAM

BY APPOINTMENT
TO H.M. QUEEN ELIZABETH II
THE QUEEN MOTHER
BRITISH LEYLAND UK LIMITED
LEYLAND CARS
MANUFACTURES OF DAIMLER, JAGUAR,
ROVER CARS AND LAND ROVERS
BIRMINGHAM AND COVENTRY

RANGE ROVER
OWNER'S MANUAL
1970-1980

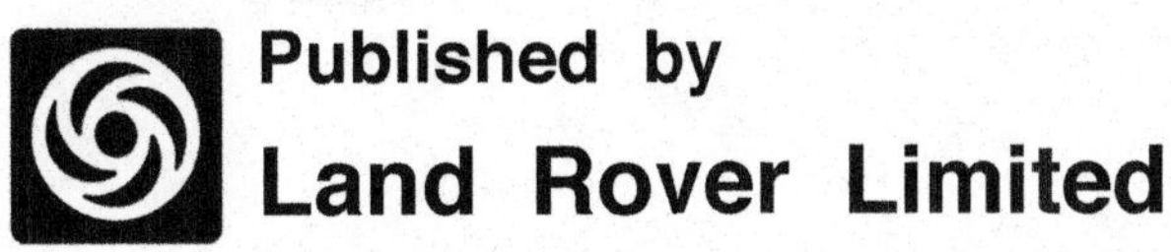

Published by
Land Rover Limited
A Managing Agent for Land Rover UK Ltd.

Publication Part No. 606917 (EDITION 2)

Land Rover Limited Service Division

LR 556

Introduction

The information in this Owner's Manual has been divided into nine sections to facilitate reference to any particular aspect of the vehicle. Section One to Five cover driving the vehicle, the use of the instruments and various controls, running requirements and a Service Guide. They should be carefully studied so you not only get the best out of the vehicle in terms of economical and efficient operation but also maximum pleasure and enjoyment when driving.

Subsequent sections include detailed instruction for the necessary maintenance and adjustment which should be carried out at regular intervals, together with notes on bulb changing, specifications of the vehicle, etc. These latter sections are for the owner who takes a personal interest in the maintenance of his or her vehicle and for other reference purposes.

On any correspondence appertaining to this vehicle the chassis number must be quoted. See page 36.

Vehicle supplied to certain territories may be fitted with additional equipment to meet legal requirements.

Information for the operation and maintenance of this equipment and any special optional equipment fitted to the vehicle is available from your Range Rover agency.

WARNING
Many liquids and other substances used in motor vehicles are poisonous and should under no circumstances be consumed and should so far as possible be kept away from open wounds.
These substances among others include anti-freeze, brake fluid, fuel, windscreen washer additives, lubricants and various adhesives.

Important Information for the Owner

Safety hints

In the interests of road safety, your attention is drawn to the following important factors:

1. The condition of the vehicle. Adherence to the routine maintenance schedules in Section Six of this book is essential in providing safe, dependable and economical motoring, also to ensure that the vehicle conforms to the various safety regulations in force.
2. Recognition of traffic and road conditions. Always observe weather and road hazards and drive accordingly.
3. Importance of using the safety harness, even for the shortest of journeys.
4. Adjustment of seat to achieve a comfortable driving position with full control over the vehicle.
5. Frequent cleaning of windscreen, rear and side windows to achieve clear vision. Use a windscreen washer solvent in the screen washer reservoir, it will assist in cleaning the front and rear screen.
6. Maintenance of correct tyre pressures. These should be checked at least each month, or more frequently when high-speed touring or under cross-country conditions, even to the extent of a daily check.
7. Maintenance of all external lights in good working order and correct setting of headlamp beams.
8. Vehicle recovery—towed procedure, see Section 1.

Ignition and steering column lock and vehicle key numbers

9. For security reasons the key numbers are not stamped on the barrel locks. Loss of the key for the ignition and steering column lock completely immobilises the vehicle. For this reason and also because the keys are of a special design obtainable only from Land Rover Limited, two ignition and steering column lock keys are supplied with each vehicle.

Owners are advised therefore in the strongest possible terms to take the following action:

(a) Immediately on receipt of the vehicle, record all the key numbers so that in case of loss, replacements can be obtained.
(b) Keep a spare ignition and steering column lock key away from the vehicle in a safe place, but where it is readily accessible.

10. The ignition and steering column lock is the latest advance in theft protection as it locks steering and ignition. Properly used it greatly reduces the hazard of theft. When leaving the vehicle, remember to remove the steering column lock key, lock the doors and tailgate.

Contents

SUPPLIED BY ....

....

....

....

Telephone No. ....

Chassis No. ....

Engine No. ....

Date of purchase ....

OWNER'S NAME ....

Address ....

....

....

Telephone No. ....

An extensive range of special optional equipment and accessories is available through your Range Rover Dealer. These range from mudflaps to winches and overdrive. In addition, a wide selection of special body conversions is also available.

The Manufacturers reserve the right to vary their specifications with or without notice, and at such times and in such manner as they think fit. Major as well as minor changes may be involved in accordance with the Manufacturers' policy of constant product improvement.

Whilst every effort is made to ensure the accuracy of the particulars contained in this Handbook, neither the Manufacturer nor the Distributor or Dealer by whom this Handbook is supplied, shall in any circumstances be held liable for any inaccuracy or the consequences thereof.

Driving Controls and Instruments

1

Paragraph numbers refer to the items illustrated, where applicable.

Front seat adjustment

1. Special safety type front seats are fitted having the safety harness secured to anchorage points on the seat. Fore and aft movement of the front seats is controlled by lifting the locking bar at the front of the seat cushion.

2. To allow easy access to the rear seat, the backrests can be tilted forwards by lifting the lever on either the inboard or outboard side of the seat.

The seats will automatically move forward on runners to provide the maximum amount of space.

Provision is made in the seat backs for the fitting of head restraints. Optional equipment.

Rear seat

3. The bench type rear seat can accommodate three people comfortably. The backrest and seat can be folded forward to provide an added loading area in the rear compartment. It is secured in the normally upright position by means of 'catches' positioned at each end of the backrest. The catches are released by a lever positioned on the rear of the backrest.

Slide the lever to the left to release the backrest.

Provision is made in the body for the fitting of safety harness (optional equipment).

Rear view mirror (interior)

4. The interior rear view mirror stem is designed to 'break out' of its spring-loaded seating if impacted. The mirror lens surround is of pliant material and the metal parts have a non-reflective black finish.

The required rear view is obtained by moving the mirror frame about its swivel; lens deflection for anti-dazzle night driving is obtained by moving the two-position spring-loaded lever, protruding from the base of the mirror.

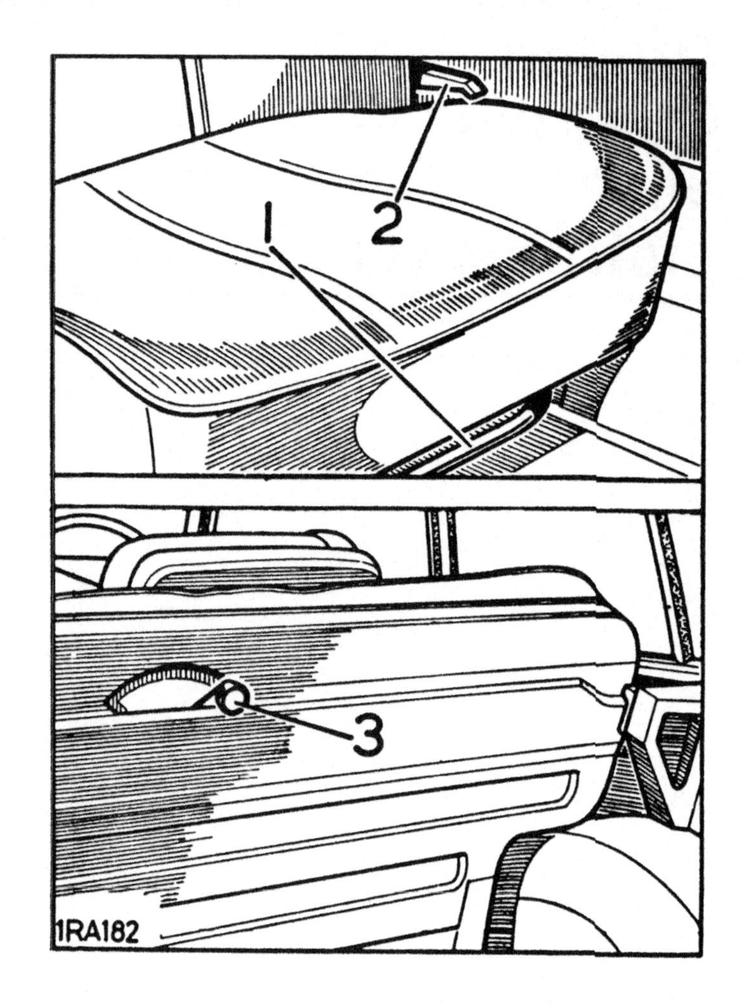

Rear view mirrors (exterior)

5. Exterior mirrors are mounted on front doors. These may be swivelled to provide the required views.

Handbrake

1. To release, pull the lever slightly back, depress the release button and push the lever down. The brakes are applied by pulling the lever back.

Steering

2. The steering will be found to be light in operation requiring only 5·55 turns of the wheel from lock to lock.

Pedals

3. Brake, clutch and accelerator pedals are the pendant type and function in the normal way. The brake and clutch operate hydraulically with servo assistance for the brakes. The accelerator pedal is connected to the carburetters by a nylon cable.

To avoid needless wear of the clutch withdrawal mechanism, do not rest the foot on the clutch pedal while driving.

Main gear lever

4. The main gear lever is used in the normal way and will engage the five gears within the range selected by the transfer lever.

Gear positions are clearly marked on the knob. To engage reverse, pull the gear lever to the right, against spring pressure. All forward gears are provided with synchromesh.

Transfer gear lever

5. The transfer gear lever is used to select the high or low range of gears; it also has a neutral (mid-way) position.

The gear lever has three positions:

(a) High range—fully rearward. In this position, the main gear lever will select the gear ratios giving normal road speeds.

(b) 'Neutral'—midway position. Used in conjunction with power-take-off equipment.

(c) Low range—fully forward. When in this position, the low range of gears will be selected by the main gear lever.

Gearbox differential lock control knob

6. The Range Rover has permanent four-wheel drive and a differential fitted in the transfer gearbox which allows a high degree of mobility in off road use.

Upon encountering conditions where traction to the road wheels becomes lost or it is obvious that traction will be lost a short distance ahead the differential can be locked by means of the differential lock control. This ensures that all road wheels obtain the maximum amount of grip.

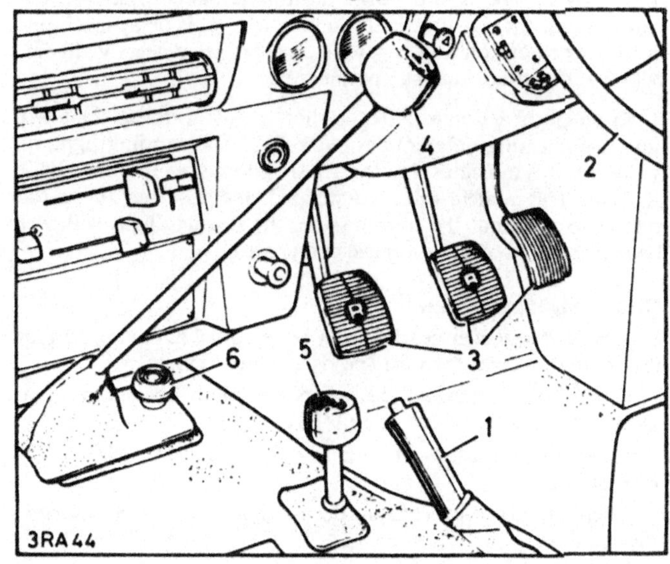

NOTE. To avoid unnecessary wear and possible damage to the transmission and final drive, it is important that wide throttle openings are not used when the vehicle is operating in 1st and 2nd gear low range with the differential locked. A return to the unlocked position must be made as soon as traction is regained.

The control knob is floor mounted, adjacent to and on the right-hand side of the main gear lever and can be engaged or disengaged at any time the vehicle is in motion, providing that the engine is running.

7. Pull knob up to engage differential lock, warning light in facia panel will be illuminated. Under certain conditions a slight delay may be experienced before the differential becomes locked, with subsequent warning light illumination. This delay is a built-in safety precaution and ensures that gears are correctly aligned before differential locking commences.

Knob pushed down. After a short delay warning light will go out indicating differential unlocked. If the warning light remains on this indicates that the transmission is 'wound up' and in tension. The vehicle must be stopped and reversed for a few metres to 'unwind' the transmission, the warning light will then be extinguished and the vehicle can proceed.

Gear changing procedure

8. The Range Rover gearbox may be regarded as having 10 gear ratios, that is eight forward speeds and two reverse.

For convenience in use these gears are evenly divided into two groups, termed 'Low' range and 'High' range.

'Low' range consists of four low forward gears, plus a low reverse gear.

'High' range consists of four normal gear ratios, plus a normal reverse gear.

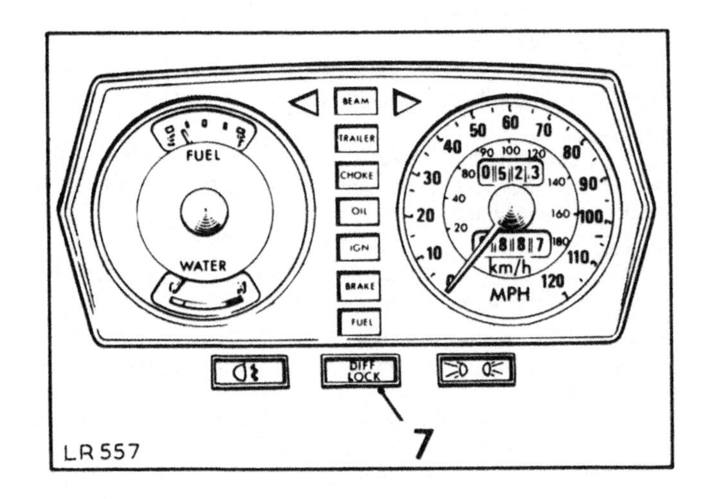

The two ranges may be used progressively when changing up, if conditions demand.

Use of gear ranges

9. As an example of how full progressive range of the gearbox may be used, consider a vehicle which is heavily laden or towing a heavy trailer and which is required to pull away from a standing start up a steep gradient. With the transfer gear in 'low' range position, the vehicle will pull away in first gear and the gear changes for the first four gears can be made in the normal way with the main gear lever. When road conditions are suitable for

high gear range they may be brought into operation without stopping the vehicle as follows:

Depress the clutch pedal. Return the main gear lever to the neutral position. Move the transfer gear lever to the neutral position. Release the clutch momentarily. Depress the clutch and move the transfer lever in 'high' position. Select second or third gear position, depending on road conditions, and release the clutch. Continue to change up in the normal way.

This operation can be carried out smoothly and quickly after a little practice. By making use of the full range of the gearbox in this manner, the clutch life will not be shortened by having to compensate for the selection of an unsuitable gear ratio.

Transfer gear changing

10. Changing from high to low, lever fully rearward to lever fully forward, should only be attempted when the vehicle is stationary. Depress the clutch pedal and push the lever fully forward; release the clutch. Should there be any hesitation in the gear engaging, do not force the lever. With the engine running, engage a gear with the main gear lever and let in the clutch momentarily; then return the main gear lever to neutral and try the transfer control again.

Vehicle recovery – towed

Since the Range Rover has permanent four wheel drive and is fitted with a steering lock it is most important, if the vehicle is to be towed, that one of the following procedures are strictly adhered to, depending on the type of tow to be undertaken.

Towing the Range Rover (on four wheels)

1. Set the main gearbox in neutral.
2. Set the transfer gearbox in neutral.
3. Set the ignition/steering lock key in position I to release the steering lock.
4.* Ensure that the differential lock is in the normal *unlocked* position.
5. Secure the towing attachment to the vehicle.
6. Release handbrake.

NOTE. Unless the engine is running brake servo cannot be maintained. This will result in a considerable increase in pedal pressure being required to apply the brakes.

***Locked differential**

If the differential is locked start the engine to provide vacuum operation to switch to unlock.

If the engine is unusable remove one of the propellor shafts.

Important. Where a front propellor shaft is to be removed check whether the four rear end fixing bolts in the gearbox flange are entered from the gearbox side. In this event they cannot readily be withdrawn. However, since the flange will revolve as soon as the vehicle is towed the four loose bolts *must* be tightly secured with nuts or suitably wired to prevent damage to the gearbox end casing.

Suspended tow by breakdown vehicle

1. Disconnect the propellor shaft from the axle to be trailed.
2. If the front axle is to be trailed it will also be necessary to set the ignition/steering lock key in position I to release the steering lock. The steering wheel and/or linkage *must* be secured in a straight ahead position.
3. The vehicle can then be raised and attached to the breakdown vehicle.

Transporting the Range Rover on a trailer

Lashing eyes are provided on the front and rear chassis members to facilitate the securing of the vehicle.

Ignition and steering column lock switch

1. The switch has four positions. Use large key.

(a) Key horizontal at '0' position. Electrical switch off. Steering column lock will be engaged during commencement of key removal. Turn steering wheel until locking plunger clicks into position. If there is difficulty in turning the key after replacement, release load on steering column lock by slightly moving the steering wheel to and fro.

Note: To prevent the steering column lock engaging it is most important that before the vehicle is moved in any way, i.e. for towing or coasting purposes, the ignition key must be inserted in the lock and turned to the position marked 'I'. If, due to an accident or electrical fault it is not considered safe to turn the key, the battery must first be disconnected.

(b) Turn right to position 'I'. Accessories can be used, that is, radio if fitted.

(c) Turn to position 'II'. Ignition and all accessories on.

(d) Continue to turn to right against spring pressure to position 'III'. Starter will operate.

(e) Should the engine fail to start at the first attempt or has stalled, it is necessary to turn the key back to the 'I' position. The operating sequence has been designed to prevent accidental locking. The key must be depressed in the 'I' position before it can be turned to the lock position '0'. The key can only be withdrawn or inserted in the lock '0' position.

Warning. If for any reason the (ignition) engine is switched off while the vehicle is in motion, **do not** attempt under any circumstances to depress, or turn the key into the lock '0' position, as this is part of the locking sequence for the steering.

To start the engine manually, engage the starting handle in the starter dog on the engine front pulley (with the handle at the bottom). The starting handle is located via a slotted hole in the front bumper bar.

Warning. To crank grasp the handle with the right hand. The thumb **must** be on the same side as the fingers. Give the handle a sharp pull **upwards** in a clockwise direction. Failure to observe this instruction may result in an injury!

Main light switch

2. The main light switch has three positions:

(a) Switch in upright position: all lamps off.

(b) Switch in centre position: side lamps on.

(c) Switch in down position: side and headlamps on.

Headlamp dipper switch, combining direction indicators, horn and headlamp flasher

3. The switch has six positions:

(a) Switch in central position: dipped headlamps.

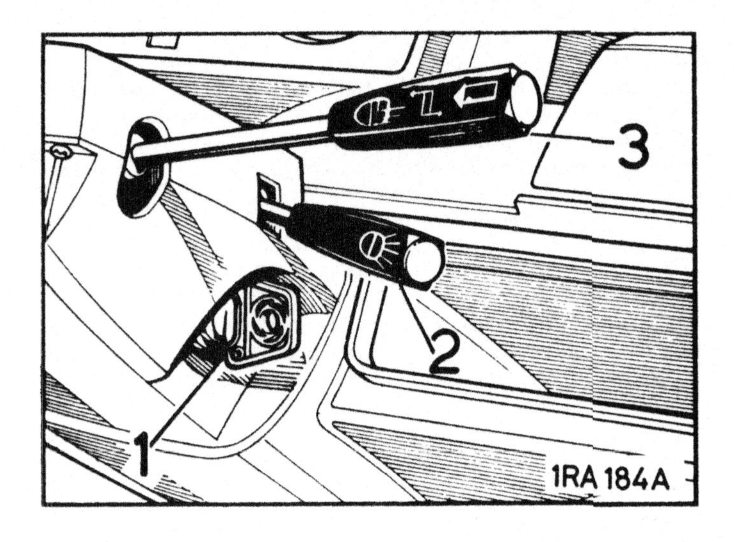

(b) Push switch fully forward: main beam.
(c) Lift fully upwards: headlamp flash. The headlamps can be flashed at any time, irrespective of other switch positions.
(d) Press dipper switch knob inwards to operate horn.
(e) Move switch anti-clockwise to indicate a left-hand turn.
(f) Move switch clockwise to indicate a right-hand turn.

Panel and instruments light switch

1. The panel and instruments light switch is operative only with the main light switch at 'side' or 'head' position.

Interior light switch

2. The interior light switch, which is immediately adjacent to the panel light switch, controls the two roof lamps, which will also be illuminated when either front door is opened.

Rear fog guard lamps switch

3. The switch has two positions, and can be operated with or without the ignition on. The rear fog guard lamps switch is operative only with the headlamps on in the dipped position.

Windscreen wiper switch and screen wash

4. The windscreen wiper switch has four positions, and is only operative when the ignition is switched on.
(a) Switch in upper position: wipers off.
(b) Switch in central position: slow-speed wiper.
(c) Switch in lower position: fast-speed wiper.
(d) Lift fully upwards, 'flick-wipe' position: wipers will operate at slow speed until switch is released.

To wash windscreen, press wiper switch knob and hold until sufficient water is on windscreen. This can be done with wiper switch on or off.

For rear screen wiper and washer switch, see following page.

Where headlamp wash/wipe facilities are fitted, these will operate in conjunction with the windscreen wiper and washer when the headlamps are illuminated.

Inspection lamp sockets

5. The sockets can be used either for a lead lamp or a trickle battery charger. The black socket is earthed.

Cold start control (illustrated on following page)

6. The cold start control has two functions:
(a) Pulled out approximately 14 to 16 mm (0.562 to 0.625 in.) increases the engine speed without mixture alteration.
(b) During the second stage of the movement the mixture is progressively enriched for cold starting. To ensure easy starting the control should initially be pulled fully out, summer and winter. After the engine has started return the control to the off position as soon as possible consistent with even running.
(c) By turning the knob slightly, the control can be locked in any position.

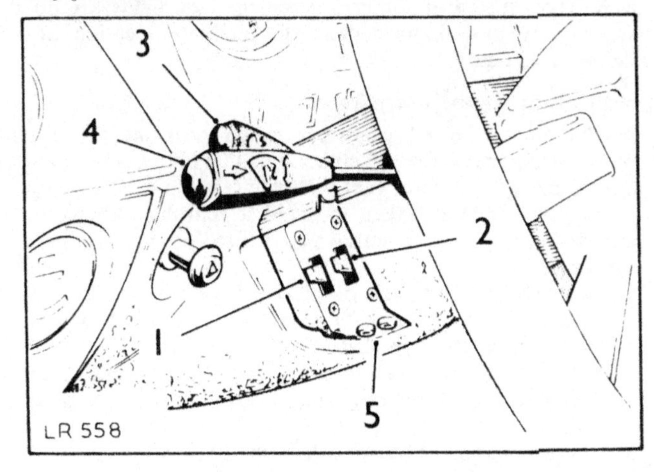

Rear screen wiper and washer switch

7. The rear screen wiper/washer switch has three positions and is only operative when the ignition is switched on.

(a) Push switch to mid-way position to operate the rear screen wiper.

(b) To wash the rear screen, push switch to lower position and hold until sufficient water is on the screen. The switch is spring loaded and will return to the mid-way position when released.

Important: Ensure that rear screen wiper is in the parked (off) position before raising the upper tailgate, otherwise damage may occur to the wiper arm and motor.

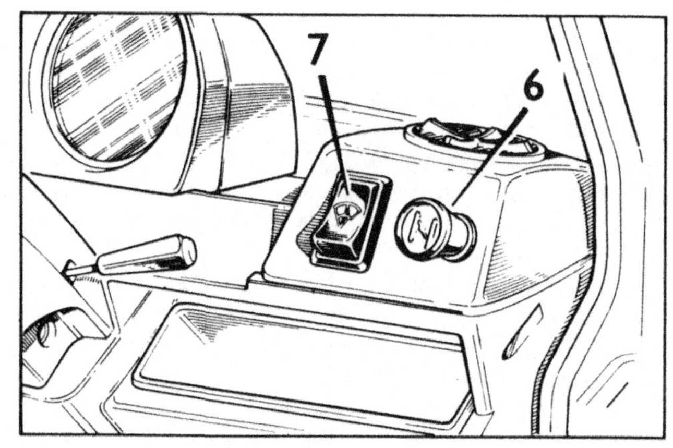

Cigar lighter

8. Operate by pressing the knob in until it clicks into position. After a few seconds it will eject to its original position when it can be withdrawn for use. When the side lights are on the socket surround is illuminated to facilitate location of the lighter in the dark.

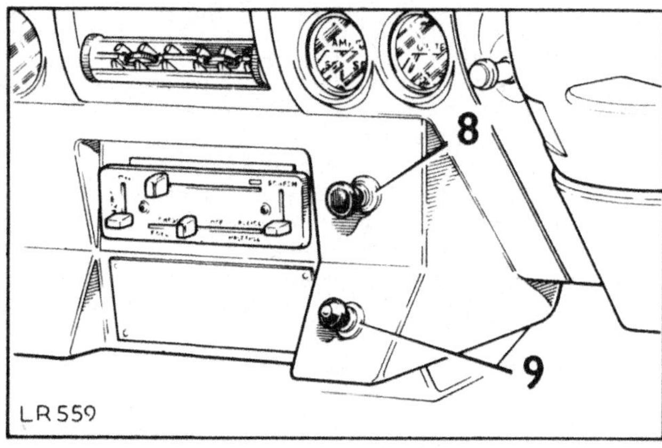

Heated rear screen, as applicable

9. Pull out the heated rear screen switch to operate the element on the rear screen. The switch knob will be illuminated acting as a reminder to the driver that the switch is in use. The following precautions must be taken to avoid irreparable damage being caused to the printed circuit which is 'fired' on to the interior of the screen.

(a) Do not remove labels or stickers from the screen with the aid of sharp instruments or similar equipment which are likely to scratch the glass.

(b) Care should be taken to avoid inadvertently scratching the glass with a ringed finger etc. when cleaning or wiping the screen.

(c) Do not clean the screen with harsh abrasives.

Ignition warning light

1. The red ignition warning light marked 'IGN' should glow when the ignition is switched on.

Brake circuit check warning light

2. The red warning light marked 'BRAKE' is most important and is arranged to warn you if there is a fluid leakage from either the front or rear braking system. If leakage occurs the warning light will come on when brakes are applied and go out when pedal pressure is released.

The warning light will also operate if a loss of vacuum occurs in the brake servo system.

Oil pressure warning light

3. The red warning light marked 'OIL' must glow when the ignition is switched on. The ignition and oil pressure lights may flicker when the engine is running at idling speed, but provided they fade out as the engine speed increases, the charging rate and oil pressure are satisfactory.

Warning lights

4. Brake circuit ignition and oil warning lights should be checked when starting the vehicle from cold; they should light up immediately the igntion is switched on and extinguish when the engine is running. If any of the above lights come on during normal running or braking, the vehicle should be stopped immediately and the cause investigated. This is of special importance in the case of the brake and oil warning lights.

Note. The ignition warning light is connected in series with the alternator field circuit. Bulb failure would prevent the alternator charging except at very high engine speeds, therefore the bulb should be checked before suspecting an alternator fault. A failed bulb should be changed with the minimum of delay otherwise the battery will become discharged.

Hazard warning light

5. When the hazard warning light switch is pulled out, all four flasher lights operate simultaneously. The red warning light [with triangular symbol] in the switch and both flasher arrows in the instrument panel will flash in conjunction with the exterior flasher lights.

Use the hazard warning system to warn following or oncoming traffic of any hazard, that is, breakdown on fast road, or an accident to your own or other vehicles.

Side lamps warning light

6. The green warning light (with symbol) will be illuminated when the side lamps are switched on.

Main beam warning light

7. The blue light marked 'BEAM' glows when the headlamp main beams are in use. Its purpose is to remind you to dip the headlamps when entering a brightly lit area, or when approaching other traffic.

Direction indicator arrows

8. The appropriate arrow flashes in conjunction with the selected set of indicator and side repeater lights. In addition the flasher unit is audible while the lights are flashing.

Should either a front, side or rear indicator bulb fail, the warning light on the side affected will remain on and the flasher unit will not be heard.

Direction indicator side repeater lamps are fitted to both front wings above the wheel arch.

Cold start warning light

9. The appearance of the amber warning light marked 'CHOKE' indicates that the engine has reached normal working temperature and that the cold start control is still fully out. The control should be returned to the normal 'in' position as soon as possible, consistent with even running.

However, the warning light will not be illuminated at the fast idle position; that is, the control out approximately 14 to 16 mm (0.562 to 0.625 in.).

Fuel level warning light

10. The green warning light will be illuminated when there is approximately 9 litres (2 gallons) left in the fuel tank. The light will remain on until the fuel supply is replenished.

Intermittent flashing may occur when cornering, etc. before the fuel level drops below two gallons.

Trailer warning light

11. The trailer warning light is only operative when a trailer is connected to the vehicle via a seven-pin socket (optional equipment). It will flash simultaneously with the vehicle indicator warning lights, thus ensuring that the trailer indicator lamps are functioning correctly. In the event of an indicator bulb failure on the trailer, the warning light will flash once only and then remain extinguished.

Differential lock warning light

12. The orange warning light will be illuminated when the gearbox differential lock control knob, located adjacent to the main gear lever, is operated. Use the differential lock only when traction to the road wheels becomes lost, i.e. under exceptional adverse cross-country conditions. A return to the off position should be made as soon as conditions permit. See pages 9 and 10 for further details.

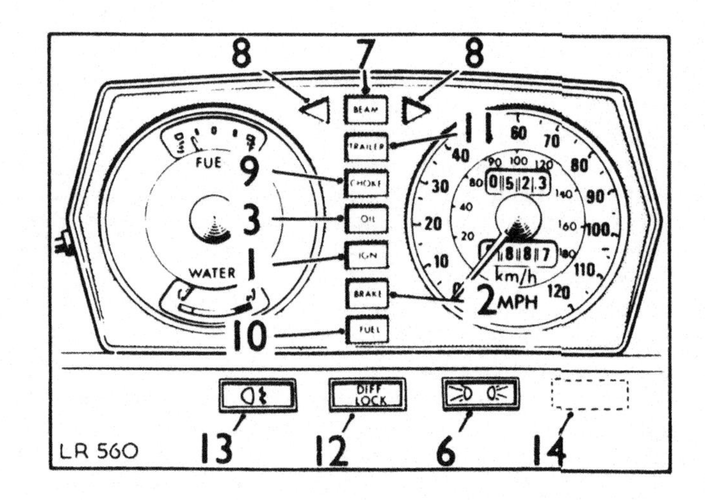

Rear fog guard lamps warning light

13. The amber warning light will be illuminated when the rear fog guard lamps are switched on.

Park brake warning light (where fitted)

14. A red warning light marked ‘Park Brake’ is illuminated when the handbrake is applied.

Speedometer

1. The speedometer incorporates total and trip mileage indicators.

Speedometer trip setting

2. Turn trip back to zero by anti-clockwise rotation of the small black knob on the instrument panel end finisher.

Fuel level indicator

3. The fuel level indicator shows the contents of the tank: the total capacity being 81,5 litres; 18 Imperial gallons; 21.5 US gallons.

Coolant temperature indicator

4. Under normal running conditions the temperature indicator needle should register in the black band. Should the needle travel to the red band during normal running, the vehicle should be stopped and the cause investigated.

Clock

5. The clock, mounted on the heater console, is an electronic type. The hands may be set by means of the black knob in the centre of the face. Push in to operate. For clock and bulb removal, see Section 7.

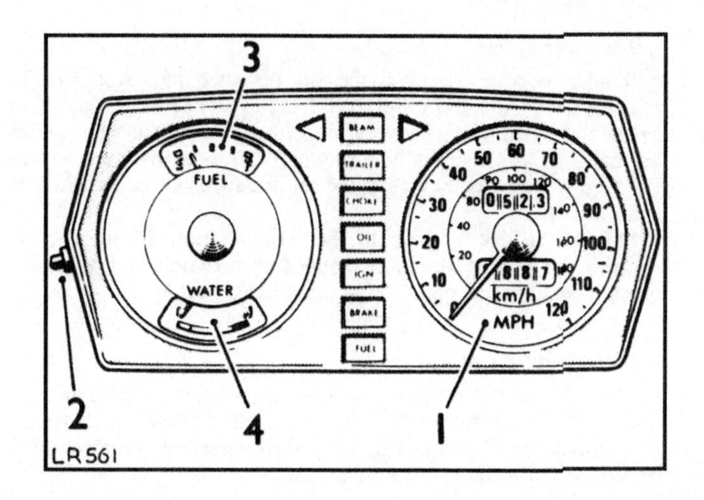

Oil pressure gauge

6. Under normal running the oil pressure indicator should register 2,1 to 2,8 kg/cm^2 (30 to 40 lb/sq in). The needle may drop below this when the engine is idling but providing the oil pressure increases to approximately 2,1 kg/cm^2 (30 lb/sq in) immediately the engine speed is increased the oil pressure can be considered satisfactory.

Should the needle drop to the zero position during normal running the vehicle should be stopped immediately and the cause investigated (see oil pressure warning light, par. 3).

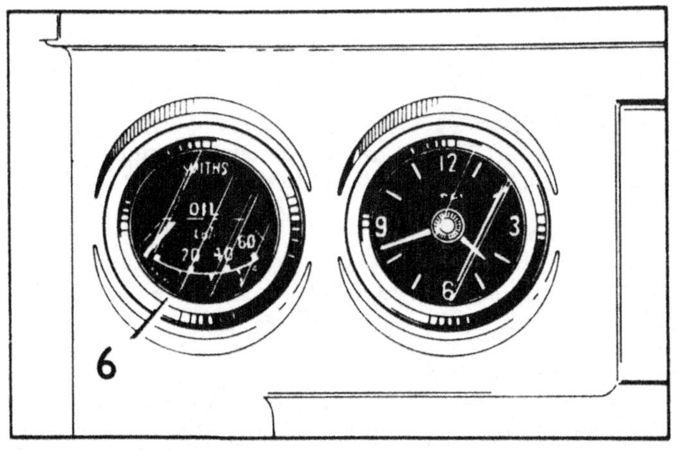

Oil temperature gauge

7. Under normal running conditions the oil temperature indicator needle should register in the white band. Should the needle travel to the (max.) red block during normal running, the vehicle should be stopped and the cause investigated.

Voltmeter

8. The voltmeter registers the battery's state of charge. With the engine running above idling speed the indicator should register within the black band 'NORMAL'. A reading above this in the high white band which continues after 10 minutes running is too high and should be investigated. A reading below this in the low white band is too low unless the headlamps and other electrical equipment are in use.

With the engine stationary but with the ignition, headlamps and other electrical equipment switched on, the indicator should register within the black band. If the reading is below this in the low white band the battery or charging system requires attention.

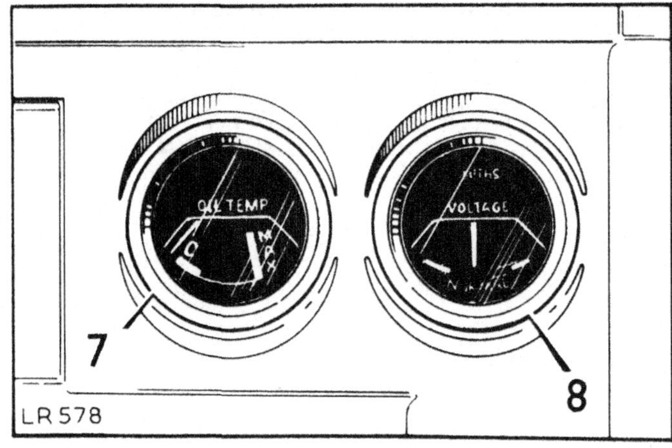

Heating and Ventilation

2

Heating

The Range Rover has a combined fresh air and recirculating heating system which has been designed so that either system can be used separately.

The recirculating heater is normally used in heavy traffic conditions to avoid obnoxious fumes entering the vehicle, also for a rapid heat build up inside the vehicle under cold conditions.

The controls are operated with the following effect:

1. Main lever. This has six positions:
(a) 'OFF'. Air entry into vehicle completely cut off.
(b) For fresh air heating: move the lever to the left to the position marked 'RAM'. Air will enter the vehicle either through the face-level louvres, if required, or through the heater to the screen, or car or both, by ram action due to the forward movement of the vehicle.
(c) Further movement to the left to the position marked 'HALF', the blower will operate at low speed.
(d) Continue movement to the left to position marked 'FULL', the blower will operate at full speed.
The blower is normally used when the vehicle speed is too low to provide sufficient heating by ram action alone.
(e) To operate the recirculating heating system, move the lever across towards position marked 'RECIRC', when either half or full speed blower can be used.

2. The top lever for temperature control.
(a) Move in blue direction to decrease heat.
(b) Move in red direction to increase heat.
(c) Action is progressive between the two.

3. Vent—the left-hand lever, which has two positions:
(a) ON: This allows entry of cool air only from the two face-level louvres on the facia panel, and also the centre face-level louvre. The face-level louvres can be adjusted rotationally and progressively to regulate the direction of air flow to the passengers.
(b) OFF: This cuts entry of air completely from the face-level louvres.

4. Screen and car—the right-hand lever, which also has two positions:
(a) 'Screen': All air is directed to the windscreen through the demister vents, either hot or cold.
(b) 'Car': All air is directed to foot level, either hot or cold, although a certain amount will continue to flow through the demister vents.

Heater distribution flap

5. With the flap closed, the main heat supply will be to the front footwells and the demister. With the flap fully open, the main heat supply will be to the rear seat passengers. With the flap in the half-way position, heat supply will be divided between the footwells and demister, and the rear seat passengers.

Side face-level louvres

6. The two side face-level louvres blow only cool air; each have knurled knob in the centre, which can be rotated to regulate the amount of air. The louvre itself is adjustable, regulating the direction of air flow.

Centre face-level louvre

7. The centre face-level louvre also blows only cool air. Open the control and increase the air flow by rotating the unit upwards. Direction of air flow to driver and passenger can be controlled by operating the centre knurled adjusters.

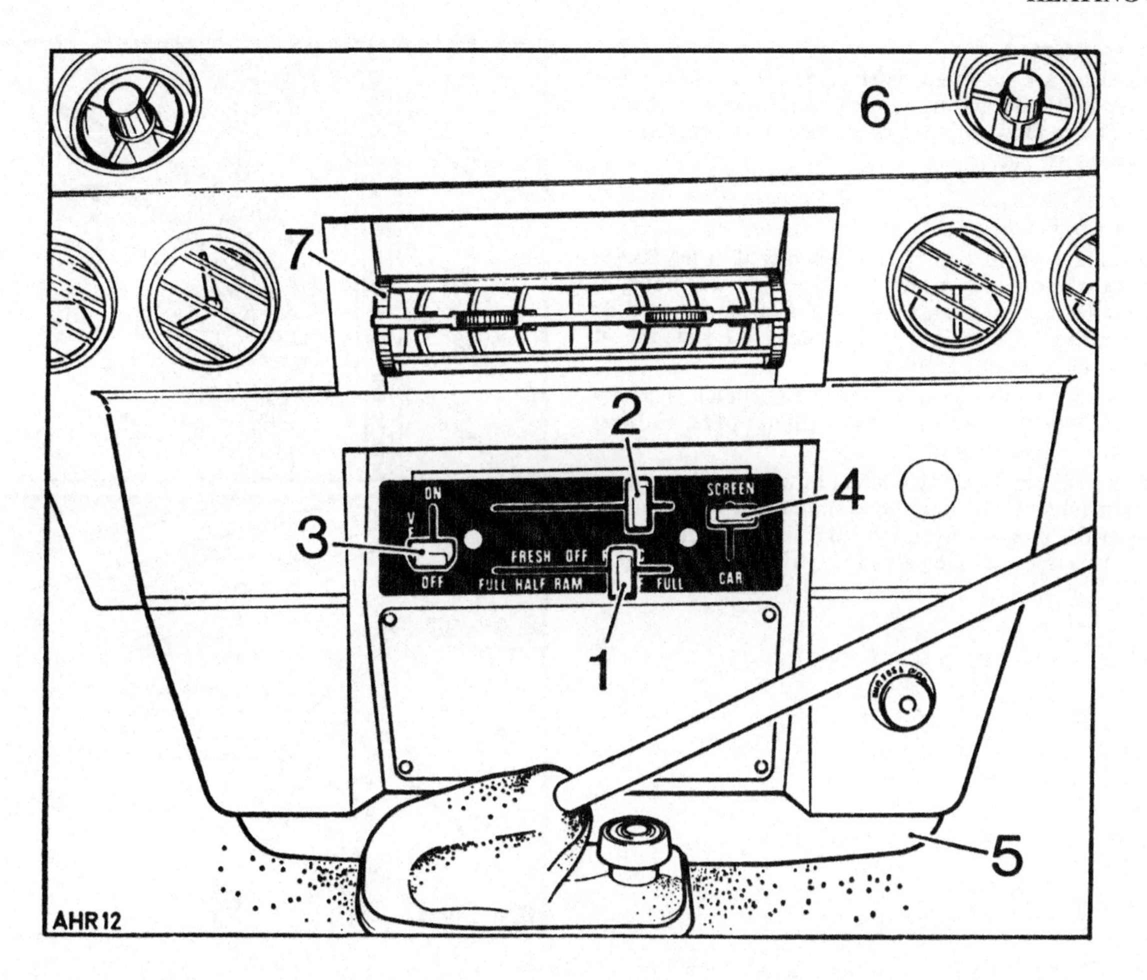
6
7
2
ON
SCREEN
4
3
FRESH OFF
OFF
FULL HALF RAM
FULL
CAR
1
5
AHR12

Door ventilator windows

1. Controlled by a flush fitting safety type catch; push in the centre button to release, and turn the catch towards the front of the vehicle. The window can then be opened to increase the air flow through the vehicle.

Through-flow ventilation

2. A through-flow ventilation system is achieved in the Range Rover by means of one-way air extraction vents incorporated in both rear quarter panels. Each vent is automatically opened or closed progressively increasing or decreasing the amount of ventilation to suit interior conditions.

Air flows round and through the interior air ventilators, and is dispensed down the inner section of the quarter panel and out through the lower exterior vent.

The system gives controlled ventilation throughout the vehicle, and the continual flow of air is also an important safety factor, as it prevents the interior becoming stuffy, and minimises any tendency to drowsiness on the part of the driver.

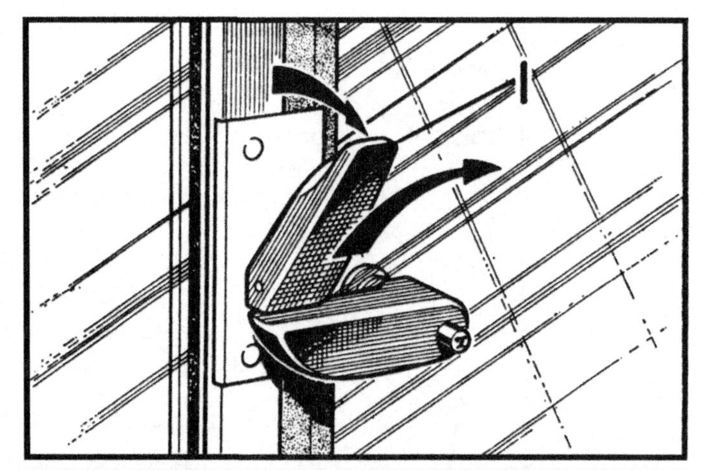

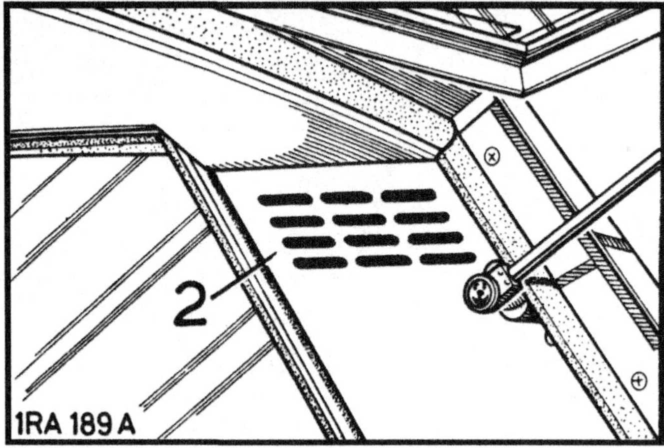

Safety Harness and Body Fittings

3

Use your seat belts

Statistics indicate that the use of safety belts can save lives. We therefore most strongly urge that the seat belts are used at all times.

Description of inertia reel type harness

Seat belts are fitted to both front seats as an integral part of the seat assembly. The belts are of the dual sensitive inertia reel type designed for one-handed engagement.

Under normal driving conditions the reel will allow the harness to extend and retract to permit normal body movement without locking but will lock automatically in the event of hard braking or fast cornering.

No adjustment is required as the automatic retraction of the of the reel retains the harness at the correct tension.

Using the harness

Using the tongue (1) draw the belt out of its slot (2) in the top of the seat back until a loop is formed, then pass the arm nearest to the door through the loop. Pass the belt across the chest and push the tongue into the buckle unit (3) attached to the inner seat fixings.

A positive click will indicate when the harness is safely locked.

Releasing the harness

To release the harness depress the button (4) on the buckle unit and allow the belt to retract gently into the slot in the top of the seat back.

Care of the belts

1. The safety belts fitted to this vehicle represent valuable and possible life saving equipment, which should be regarded with the same importance as steering and brake systems. Frequent inspection is advised to ensure continued effectiveness in the event of an accident.
2. Inspect the belt webbing periodically for signs of abrasion or wear, paying particular attention to the fixing points. Do not attempt to make any alternations or additions to the seat belts or their fixings as this could impair their efficiency.
3. If worn and correctly stowed, deterioration will be kept to a minimum and protection to a maximum.
4. Seat belt assemblies must be replaced if the vehicle has been involved in an accident, or if upon inspection, there is evidence of cutting or fraying of the webbing, incorrect buckle or tongue locking function; and/or any damage to the buckle stalk cabling.

Harness cleaning

Do not attempt to bleach the belt webbing or re-dye it. If the belts become soiled, sponge with warm water using a non-detergent soap and allow them to dry naturally. Do not use caustic soap, chemical cleaners or detergents for cleaning; do not dry with artificial heat or by direct exposure to the sun.

Checking inertia reel mechanism

The following road test must be carried out only under maximum safe road conditions, i.e. on a dry, straight, traffic free road.

With the safety harness fitted, drive the car at 8 km/h (5 mph) and brake sharply. The safety harness should lock automatically, holding the user securely in position.

Snatch test

Whilst seated, fasten the seat belt and grip the shoulder belt at approximately shoulder level with the opposite hand. Pull the belt sharply in a downwards direction, the belt should lock.

It is important when braking that the body is not thrown forward in anticipation.

Rear seat belts (optional equipment)

Provision is made to fit two *outer* inertia reel type harnesses and a *centre* static (lap) type harness to the rear vehicle body.

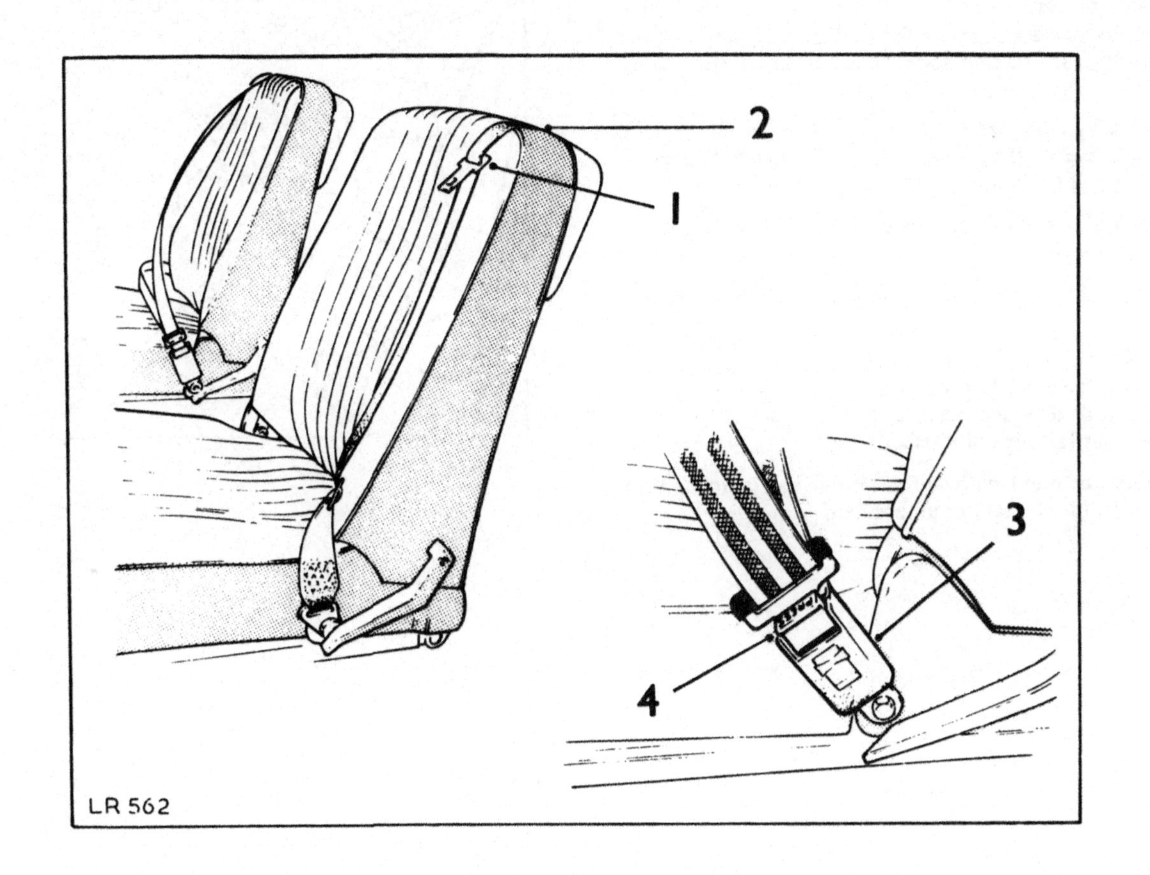
2
1
3
4
LR 562

Doors and lock controls

1. The front doors are extra wide for ease of entry and exit for rear seat passengers. The doors are locked from the outside by the ignition key.

2. Both doors can be opened from inside by front or rear passengers via twin release handles. Both handles operate independently; pull to release.

3. The door pull handles are located immediately above the door release handles.

4. The exterior door handle is operated by pulling outwards.

5. As a safety precaution an interior catch is provided on both doors to prevent the doors from being opened accidentally from the inside. The catch has two positions:
(a) Fully forward: door locked.
(b) Fully rearward: door unlocked.

6. Both doors are fitted with an anti-burst device to prevent them flying open in the event of an accident.

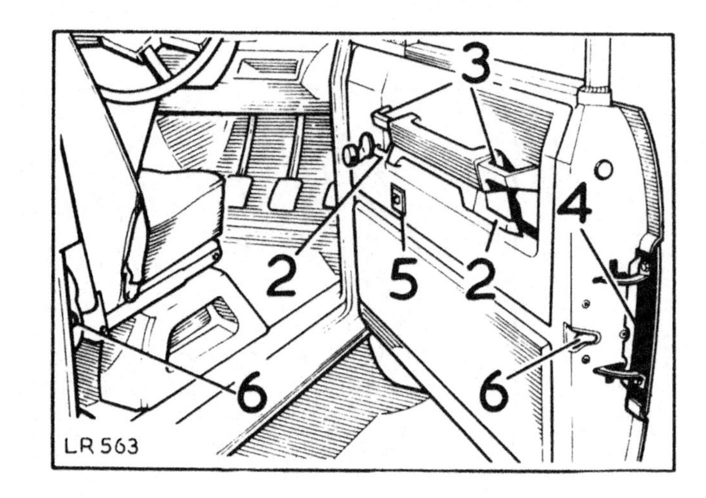

Bonnet lock control

1. To open the bonnet, pull the control knob, located below the glove box, driver's side. This disengages the locking plate and allows the bonnet to spring open sufficiently to insert the fingers.

2. The bonnet has a safety catch which must also be released by lifting the bonnet slightly and pulling forward the catch at the front right-hand side of the bonnet.

3. In the fully open position the bonnet can be supported by the prop rod, which should be engaged in the slotted hole in the top of the radiator grille.

4. To close the bonnet, replace the prop rod, lower bonnet to about 304 mm (12 inches) above the grille and allow to drop into position.

5. If it is necessary to push down on the bonnet, do this with the palms of both hands at the front edge.

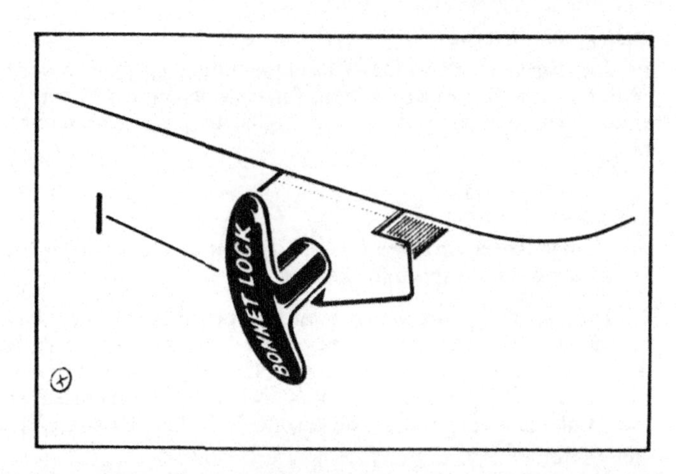

Sliding side windows

1. The large rear side windows are of the sliding type, the forward section having five open positions for varying degrees of ventilation. Each sliding window is controlled by a simply-operated catch.

Tailgates

The Range Rover provides full width upper and lower tailgates, allowing maximum space for loading, etc.

2. The upper tailgate, which must be opened before the lower tailgate can be lowered, is released by depressing the locking button in the centre of the handle. The tailgate can then be raised to its fully elevated position, where it is supported by means of telescopic rods. The tailgate is locked by using the square-headed key.

Important: Ensure that the rear screen wiper is in the parked (off) position before raising the upper tailgate, otherwise damage may occur to the wiper arm and motor.

3. The lower tailgate is an all-steel construction for greater strength. It is supported in the lowered position by means of folding stays and has a single centre locking handle above the hinged number plate plinth. The handle has three positions:

(a) Handle in fully left position: tailgate locked.
(b) Handle in centre position: tailgate catch.
(c) Handle in fully right position: tailgate released.

Spare wheel

4. The spare wheel is mounted in the interior of the vehicle, positioned at the left-hand rear side. A fabric wheel cover is fitted over the wheel to prevent soiling of articles in the vehicle.

Fuel filler

1. A lockable fuel filler cap is located on the rear right hand wing and is locked with the glove box key.

When the cap is unlocked it will automatically hinge open.

Glovebox

2. A lockable glovebox is provided on the passenger's side and can be opened by pulling the release handle rearwards. Lock with the square-headed key.

Facia panel

3. The facia panel has a 'grained' finish, with a centrally-placed radio speaker grille, a tray with a non-slip rubber mat is provided on the passenger's side.

4. A map pocket, which alternatively can be used to accommodate a radio, is located on the driver's side adjacent to the steering column.

Body care

5. It is always preferable to clean the bodywork and exterior trim with water and sponge, using plenty of water. Dry with a chamois leather, polish occasionally using any good brand of wax car polish.

The use of salt on the roads during frosty weather, sometimes in quite strong concentrations, is widely practised. Due to its highly corrosive nature, salt deposited should be washed off as soon as possible by thorough underwashing of the vehicle with a hose.

It is most important that detergents are not used when cleaning seats, etc. Use a damp cloth or soft brush with a little mild soap.

'NYLON' seat trim

Clean off surface dirt and dust using a soft brush. Wash with warm water and mild soap. Allow to dry naturally.

Vinyl covered rear quarter panels

Wash the vinyl surface over with warm soapy water (use soap flakes or mild tablet soap). If dirt is ingrained the use of a soft nail brush will help. Rinse off with clean cold water ensuring that all soap is removed. During normal cleaning of the car the vinyl will not be affected by mild detergents such as are used in Automobile Car Washes. Avoid the use of wax polish, creams, solvents or strong detergents. Under no circumstances should White Spirit or Petrol be used to remove oil or grease marks from the vinyl surface.

Running Requirements and Recommended Lubricants

4

Fully illustrated details of all maintenance required will be found in Section 6 of this manual, but you should note the following:

Running-in period

1. Progressive running-in of your new Range Rover is important and has a direct bearing on durability and smooth running throughout its life.

The most important point is not to hold the vehicle on large throttle opening for any sustained periods. To start with, the maximum speed should be limited to 80 to 95 kph (50 to 60 mph) on a light throttle and this may be progressively increased over the first 2.500 km (1,500 miles).

Coolant

2. The water level should be checked periodically.

Warning. Do not remove the expansion tank filler cap when the engine is hot, because the cooling system is pressurised and personal scalding could result.

When the engine is cool, remove the expansion tank filler cap by first turning it anti-clockwise a quarter of a turn, and allow all pressure to escape, then turn it further in the same direction to lift off. When replacing the filler cap it is important that it is tightened down fully, not just to the first stop. Failure to tighten the filler cap properly may result in water loss, with possible damage to the engine through overheating.

With the engine cold the correct coolant level should be up to the 'Water Level' plate, located inside the expansion tank just below the filler neck.

Never top-up with water only, always use a solution of water and the correct type of anti-freeze or water and inhibitor. See lubrication chart at the end of Section 4.

Do not forget to keep the screenwasher bottle filled with water; add 'Isopropyl Alcohol' or methylated spirits in the winter to prevent freezing.

Frost precautions

3. As it is impracticable to drain the cooling system fully due to water being retained in the heating system, a special anti-freeze mixture is used in the Range Rover during the winter and summer months. Anti-freeze mixture is also used to prevent corrosion of the aluminium alloy engine parts. It is most important therefore if the cooling system is drained or topped up at any time either winter or summer, to refill with a solution of water and the correct type of anti-freeze or water and inhibitor, where anti-frost precautions are not necessary, otherwise damage to the engine will result.

Battery acid level

4. Make sure that the battery acid level is above the top of the separators in each cell. Do not over-fill.

Tyres

5. The 205R x 16 radial-ply tyres used on Range Rover models have been specially selected to give the best all-round performance.

Warning: Many off-road types of tyre have a maximum speed capability below that of the Range Rover therefore when tyre replacements are required radial ply tyres of an approved type must be used. Under no circumstances should cross-ply tyres be used as replacements. Consult your local Rover Distributor or Dealer for guidance if in any doubt concerning the type of tyre required.

Tyre pressures

6. These should be checked at least every month for normal road use and at least weekly, preferably daily, if the vehicle is used off the road. See General Data, Section 9.

Tyre wear

7. It is illegal in the UK and many other countries to continue to use tyres with excessively worn tread. Tyre wear should be checked at every maintenance inspection. See details in 'Routine maintenance and adjustments – Exteriors', Section 6, page 50.

Fuel recommendations

1. The Range Rover engine has been designed to run on 91-93 octane fuels. No advantage will be gained by the use of higher octane fuels.

Brakes

2. Never coast downhill with the engine switched off as the brake servo will not be operative. The brakes will however function through the hydraulic system when the brake pedal is depressed, but more foot pressure will be required.

Tools

3. The jack, together with a tool roll, will be found attached to the rear off-side body panel. A jack extension with a wooden handle and a starting handle are also provided.

IMPORTANT POINTS TO REMEMBER

1. Read the Rover Service Guide, Section Five of this book, which contains important information for the Owner.
2. If spark plug replacements are required, use only the correct type as specified in the Data Section of this book. However, should emergency compel, the nearest alternative fuel or spark plugs may be used temporarily, subject to a speed restriction of 96 kph (60 mph).
3. Maintain correct tyre pressures.
4. Let a Rover Distributor or Dealer service your Range Rover and use only genuine Rover parts.

Recommended lubricants, fuel and fluids – UK

1. Use only the recommended grades of oil as set out below.
The oil level dipstick will be found on the left-hand side of the engine and the oil filler cap is screwed into the right-hand rocker cover at the front of the engine.

2. Oil consumption is likely to improve during the first 6.000 km (4.000 miles) of the vehicle's life as the piston rings, etc. bed in.

These recommendations apply to temperate climates where operational temperatures may vary between —10°C (14°F) and 32°C (90°F).

COMPONENT	BP	CASTROL	DUCKHAMS	ESSO	MOBIL	PETROFINA	SHELL	TEXACO
Engine and carburettor dashpots * **Oils must meet BL Cars specification BLS-OL-02 or the requirements of the CCMC**	BP Super Visco-Static 20-50	Castrol GTX	Duckhams Q Motor Oil	Esso Uniflo 15W/50	Mobil Super 15W/50	Fina Supergrade Motor Oil 20W/50	Shell Super Multigrade 20W/50	Havoline Motor Oil 20W/50
Main gearbox, overdrive, transfer gearbox *	BP Super Visco-static 20-50	Castrol GTX	Duckhams Q Motor Oil	Esso Uniflo 15W/50	Mobil Super 15W/50	Fina Supergrade Motor Oil 20W/50	Shell Super Multigrade 20W/50	Havoline Motor Oil 20W/50
Final drive units **Swivel pin housings** **Steering box**	BP Gear Oil SAE 80EP	Castrol Hypoy Light	Duckhams Hypoid 80	Esso Gear Oil GP 80W	Mobil Mobilube HD 80	Fina Pontonic MP SAE 80	Shell Spirax 80 EP	Texaco Multigear Lubricant EP 80
Power steering fluid reservoir, as applicable	BP Autran B	Castrol TQF	Duckhams Q-Matic	Essoglide	Mobil ATF 210	Pursimatic 33F	Shell Donax T7	Texamatic Type F
Lubrication nipples **(ball joints, hubs, propshafts)**	BP Energrease L2	Castrol LM Grease	Duckhams LB 10	Esso Multi-Purpose Grease H	Mobilgrease MP or MS	Fina Marson HTL 2	Shell Retinax A	Marfak All-purpose Grease
FUEL AND FLUIDS **Fuel**	91 to 93 Research Octane Fuel, 2-star grade in the United Kingdom with standard ignition timing † 85-91 Research Octane Fuel, with reset ignition timing †							
Windscreen Washers	Unipart All Seasons Screen Washer Fluid							
Brakes and Clutch reservoirs	Unipart Universal Brake Fluid or other brake fluids having a minimum boiling point of 260°C (500°F) and complying with FMVSS 116 DOT3 or SAE J1703 specification.							
Engine cooling system	Unipart Universal Anti-freeze.							
Inhibitor solution for engine cooling system	Marston Lubricants SQ36—Coolant inhibitor concentrate. For summer use only when frost precautions are not necessary.							

* Unipart Super Multigrade Motor Oil is recommended for these applications.
† See General Data—engine details for alternative ignition settings.

	Service Classification		Ambient Temperature °C
	Performance Level	SAE Viscosity	-30 -20 -10 0 10 20 30
ENGINE AND CARBURETTER DASHPOTS	Unipart Super Multi-grade Motor Oil or other oils conforming to BL Ltd. SPECIFICATION BLS-OL-02 or the requirements of CCMC or API-SE.	5W/20	
		5W/30	
		5W/40	
		10W/30	
		10W/40	
		10W/50	
		15W/40	
		15W/50	
		20W/40	
		20W/50	
MAIN GEARBOX, OVERDRIVE AND TRANSFER GEARBOXES	BLS-OL-02 *or* to requirements of CCMV or API-SE	20W/50	
FINAL DRIVE UNITS SWIVEL PIN HOUSINGS STEERING BOX	MIL-L-2105A	Hypoid 80	
POWER STEERING	ATF Type F or 6		
LUBRICATION NIPPLES (HUBS, BALL JOINTS etc.)	NLGI—2 Multipurpose Grease		
BRAKE & CLUTCH RESERVOIRS	Unipart Universal Brake Fluid or other Brake Fluids having a minimum boiling point of 260°C (500°F) and complying with FMVSS DOT 3 or SAE J1703.		
ENGINE COOLING SYSTEM	Unipart Universal Antifreeze. Where frost precautions are not necessary use Marston Lubricant SQ36 to prevent corrosion of the engine alloy.		
WINDSCREEN WASHERS	Unipart All-Seasons Screen Washer Fluid.		
AIR CONDITIONING SYSTEM Refrigerant Compressor Oil	METHYLCHLORIDE REFRIGERANTS MUST NOT BE USED. Use only with refrigerant 12. This includes 'Freon 12' and 'Arcton 12'. Shell Clavus 33 BP Energol LPT 100 Sunisco 5 Texaco Capella B		

Vehicle identification number (VIN)

The plate carrying the VIN together with the recommended maximum vehicle weights will be found under the bonnet rivetted to the panel above the front grille.

The VIN is also stamped on the right side of the chassis adjacent to the front shock absorber. Always quote the complete number (B) when writing to the Company or your Distributor or Dealer on any matter concerning your Range Rover.

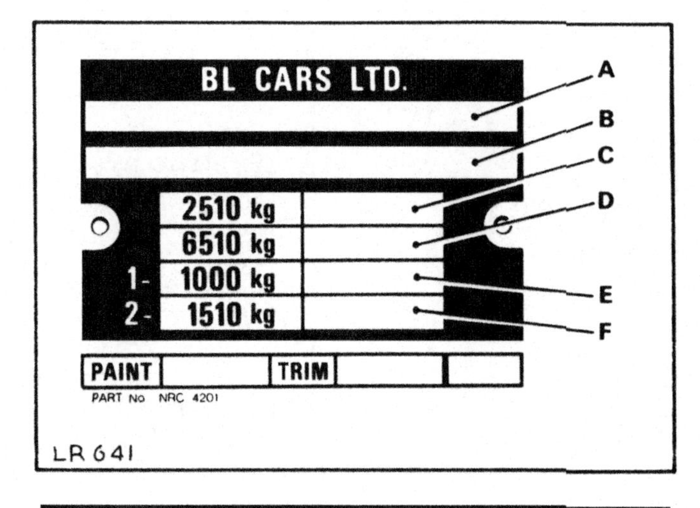

LR 641

Key to vehicle identification number plate

A Type approval
B VIN (maximum of 14 digits)
C Maximum permitted laden weight
D Maximum vehicle and trailer weight
E Maximum road weight – front axle
F Maximum road weight – rear axle

Engine serial number

2. The engine serial number is stamped on a cast pad on the cylinder block between numbers 3 and 5 cylinders.

Do not quote this number unless requested.

NOTE: The compression ratio of the engine is also given above the engine serial number.

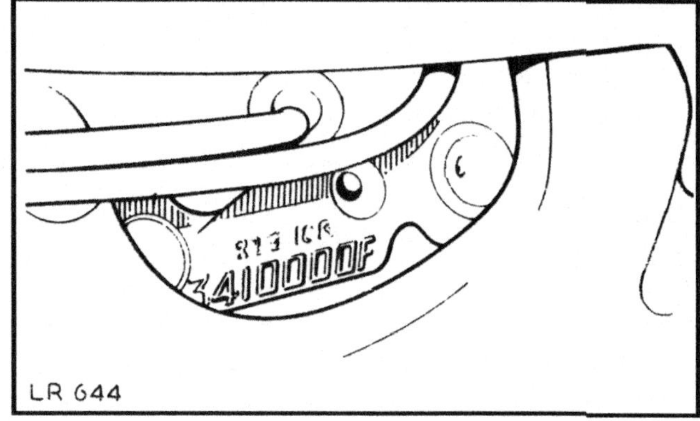
LR 644

Rover Service Guide

5

. Land Rover Limited sets high standards in the design, pecification and production of its vehicles and desires that these hould give reliable and satisfactory performance.

It is therefore strongly recommended that owners and users of Rover vehicles should familiarise themselves with the following nformation which is issued for the specific purpose of helping hem and which is set out under the following headings.

The new vehicle
Maintenance attention
General notes
Owner's Service Statement

he new vehicle

. With every new vehicle special literature is provided and it is f importance that this should be made full use of. The literature onsists of the following:

(i) Owner's Manual: This book, which gives general information about the vehicle, also incorporates notes on service, the Owner's Service Statement and full information on how to carry out the necessary maintenance.

i) Passport to Service which gives details of maintenance required and also includes spaces for the Distributor or Dealer to sign and stamp to certify that the work has been carried out at the appropriate intervals.

The operations carried out by your Distributor or Dealer will e in accordance with current recommendations and may be ubject to revision from time to time.

Upon receiving the new vehicle, the owner should immediately:

(i) Examine the Owner's Manual for advice on new features and as an aid to getting the best out of the vehicle.

ii) Arrange with a Rover Distributor or Dealer to carry out regular maintenance attention.

laintenance attention

. Efficient maintenance is one of the biggest factors in ensuring continuing reliability and efficiency. For this reason detailed schedules have been prepared so that at the appropriate mileages owners may know what is required.

(i) The Pre-delivery Inspection is a very important first step in the work of preventative maintenance. The Dealer responsible for the sale of the vehicle will have completed the work involved. There is provision in the Passport to Service for certification that this work has been carried out.

(ii) The Free Service Inspection should be carried out by the Dealer responsible for the sale of the vehicle to the owner at or about 1.500 km (1,000 miles). A charge is made only for the lubricants, etc. used in carrying out the service. Where for any reason it is not convenient for this free service to be carried out by the Dealer responsible for the sale, it can, by prior arrangement with such Dealer, be carried out by any other Rover Distributor or Dealer.

(iii) The Maintenance Schedules are based upon intervals of 5.000 km (3,000 miles) or 3 months.
These Maintenance Schedules are not priced but guidance is given to Dealers upon the actual time required to carry these out.

GENERAL NOTES

Service parts and accessories

4. **Genuine ROVER and UNIPART parts and accessories are designed and tested for your vehicle and have the full backing of the Owner's Service Statement. ONLY WHEN GENUINE ROVER AND UNIPART PARTS ARE USED CAN RESPONSIBILITY BE CONSIDERED UNDER THE TERMS OF THE STATEMENT.**

In accordance with the Company's policy of continued improvement, new parts will be introduced regularly into the UNIPART range and they should be used when servicing or replacing parts on your car.

For more information on UNIPART see your Rover Distributor or Dealer.

Genuine ROVER and UNIPART parts and accessories are supplied in cartons and packs bearing either or both of these symbols.

Safety features embodied in the vehicle may be impaired if other than genuine parts are fitted. In certain territories, legislation prohibits the fitting of parts not to the vehicle manufacturer's specification. Owners purchasing accessories while travelling abroad should ensure that the accessory and its fitted location on the vehicle conform to mandatory requirements existing in their country of origin.

Labour charges

5. The company does not issue detailed schedules of repair charges but guidance is given to Dealers on the normal times required for the majority of repair and maintenance operations (not to accidental damage to bodywork, etc.).

Over the last few years service labour costs have risen considerably and where a high standard of work is looked for, the higher price of labour charges is inevitable.

Owner's Service Statement (Warranty)

6. Land Rover Limited issues under the heading of Owner's Service Statement an undertaking regarding its Service policy.

Home market: The Owner's Service Statement is supplied in the Literature Pack.

Export markets: The Warranty (Owner's Service Statement) should be obtained from the Distributor or Dealer at the time of purchase.

The following notes are given for guidance in the event of a claim being put forward:

(i) The vehicle or the part in respect of which a claim is made must be taken immediately to a Rover Distributor or Dealer. This should, wherever possible, be the Distributor or Dealer responsible for the sale of the vehicle to the owner.

(ii) The Distributor or Dealer will examine the parts or vehicle and will without charge advise on the action to be taken in respect of the claim. It will be noted that the Company must reserve the right to examine any alleged defective parts or material should they think fit prior to the settlement of any claim.

(iii) It must be understood that the factors of wear and tear and any possible lack of maintenance or unapproved alteration will be taken into consideration in respect of any claim submitted.

(iv) It will be noted that tyres and glass are expressly excluded. The manufacturers of those tyres which the Company fits as standard to its vehicles will always be prepared to consider any genuine claim.

(v) Where this has not already been done, it is recommended that owners should arrange with their Insurance Company to provide separate cover for the glass at the small extra cost involved.

Routine Maintenance and Adjustments

Emission control

As air pollution from all sources is increasing, new and more stringent regulations are continually being introduced to limit the amount of harmful emissions from the internal combustion engine.

This requirement therefore determines the specification and type of equipment fitted to the vehicle and also the calibration requirements, for such equipment.

Owners should ensure that whenever new parts are fitted to their vehicles they obtain from the Distributor or Dealer who has carried out the repairs assurance in writing that the parts concerned conform to the safety and emission control regulations currently in force.

Range Rover models supplied to European countries where emission control regulations apply are specially equipped to control the emissions of hydrocarbons and carbon monoxide from the exhaust system.

On the Range Rover V8 engine crankcase emission control is achieved by venting the crankcase fumes to the intake manifold to be burnt in the combustion chambers.

Exhaust emissions are controlled by alterations to carburation characteristics and ignition settings. Carburetter adjustments and ignition timing are accurately set at the factory and under normal circumstances do not require attention except at the specified maintenance periods as detailed on the following pages.

However should it become necessary to check any aspect ot carburetter adjustment or ignition timing, the work must be carried out by a Rover Distributor or Dealer who has the specialised equipment needed to carry out adjustments to the close limits necessary to ensure that the engine conforms to legal requirements in respect of exhaust emission.

Safety features

The more important of the safety features incorporated in the Range Rover are detailed below:

(a) **Brakes**

The Range Rover has primary and secondary braking, consisting of a dual line system, the front brakes having four pistons in two pairs, one pair piped separately, being the secondary system, the other pair combined with the rear brakes, being the primary system, and so designed to function should there be a failure to one or more component parts.

For instance, should the front secondary system fail, half the front brakes plus rear brakes will continue to operate. Conversely braking power will still be available if a component on the primary system fails, in which case half the front brakes will operate.

(b) **Steering**

Bulkhead located steering unit with jointed steering column ensures minimum direct penetration towards the driver if the vehicle is involved in an accident.

(c) **Seats**

High strength safety seats are fitted, with built-in lap and diagonal safety harness.

(d) **Door lock anti-burst**

Both front doors are fitted with an anti-burst device to prevent the doors flying open in the event of an accident.

(e) **Interior mirror**

Interior mirror with lens deflection for anti-dazzle night driving. The mirror stem is designed to 'break-out' of its spring-loaded seating if impacted.

(f) **Hazard warning system**

A switch operates all four flashers simultaneously; use the system to warn following or oncoming traffic of any hazard, that is, breakdown on fast road, or an accident to your own or other vehicles.

Notes on general maintenance

This Section has been prepared to give clear and simple information necessary for the efficient care and maintenance of your vehicle.

Lubrication and regular service maintenance are necessary to keep any vehicle in good mechanical condition and to minimise emissions during normal driving. All the items which require regular maintenance are shown in this Section in terms of mileage which would apply in a temperate climate. Climatic and operating conditions affect maintenance intervals to a large extent; in many cases, therefore, the determination of such intervals must be left to the good judgement of the owner or to advice from a Rover Distributor or Dealer, but the recommendations will serve as a firm basis for maintenance work.

Of particular importance are the following items:

IMPORTANT

1. Check engine oil level and water level in radiator daily or weekly depending on operating conditions.

2. Drain and refill engine sump every 10.000 km (6,000 miles) or every six months, whichever comes first.

3. Tyre pressures. Check at least every month for normal road use and at least weekly, preferably daily, if the vehicle is used off the road. (See General Data, Section 9).

4. Tyre wear. It is illegal in the UK and many other countries to continue to use tyres with excessively worn tread. Tyre wear should be checked at every maintenance inspection. See details in 'Routine maintenance and adjustments – Exterior', Section 6, page 50.

5. Every month and every maintenance inspection check fluid level in brake fluid reservoir and battery acid level.

6. Brakes. Change brake fluid every 30.000 km (18,000 miles) or eighteen months. The fluid should also be changed before touring in mountainous areas if not done in the previous nine months. Use only Unipart Universal Brake Fluid or other brake fluids having a minimum boiling point of 260°C (500°F) and complying with FMVSS 116 DOT3 or SAE J1703 specification, from sealed tins.

   Renew all rubber seals in the complete brake system and all hydraulic hoses every 60.000 km /36,000 miles) or 3 years. Drain the brake fluid reservoir and flush the system. Refill with the correct fluid.

7. Owners are under a legal obligation to maintain all exterior lights in good working order; this also applies to headlamp beam setting, which should be checked at regular intervals by your Rover Distributor or Dealer.

Fuel recommendations

The engine is designed to run on 91-93 research octane fuel, two-star grade in the United Kingdom.

Engine

Under adverse conditions, such as driving over dusty roads or where short stop-start runs are made, oil changes, attention to the engine flame traps and breather filter replacement must be more frequent.

Air cleaner

When the car is driven over dusty roads the elements should be changed more frequently.

Propeller shaft

Under tropical or sandy and dusty conditions, the sliding joint must be lubricated frequently to prevent ingress of abrasive materials.

Lubricants

The lubrication systems of your vehicle are filled with a high quality oil. You should always use a high quality oil of the correct viscosity range in the engine, gearboxes and axles during subsequent maintenance operations or when topping up. The use of oils not to the recommended specification can lead to high oil and fuel consumption, excessive wear and ultimately in damage to the engine, gearbox or axle components.

Oils to the recommended specification made by reputable manufacturers contain additives which disperse the corrosive acids formed by combustion and also prevent the formation of sludge which can block oilways. Additional additives should not be used.

1. On the following pages will be found full instructions on how to carry out the maintenance and adjustments required on the Range Rover models.

2. The sequence of operations under the headings of Passenger Compartment, Exterior, Underbody, Engine Compartment and Road Test will enable the work to be carried out in the most efficient manner.

3. Absolute cleanliness is essential when carrying out maintenance.

4. Throughout this section will be found a number of layout illustrations; these show the position of the various parts which require maintenance and are designed to give the owner assistance in quickly locating the items concerned.

In each case the numbers on the illustrations refer to the fig. numbers on the detail drawings that follow.

The numbers on the detail drawings refer to the paragraph numbers of the text.

Range Rover maintenance location points for:

1 Headlamp unit adjustment
2 Wiper blade replacement
3 Front brake pads
4 Rear brake pads
5 Wheel alignment

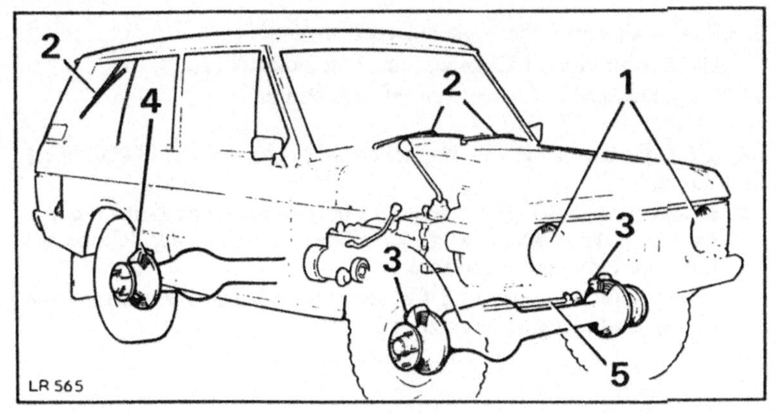

Steering—Every maintenance inspection.

1. Check steering wheel backlash—zero in straight ahead position. See your Distributor/Dealer if rectification is necessary.

Foot and handbrake—Every 5.000 km (3,000 miles) or 3 months.

2. Check operation of foot and handbrake, ensure that the brake pedal travel is not excessive and maintains a satisfactory pressure under normal working load. Excessive pedal travel indicates worn brake pads.

   If the brakes feel spongy, this may be caused by air in the hydraulic system and this must be removed by bleeding the system at the wheel disc cylinders. See page 95.

   Prior to this operation all hydraulic hoses, pipes and connections should be checked for leaks and any leaks rectified.

   Check operation of handbrake, ensure that it holds the vehicle satisfactorily. If adjustment is required see page 59.

Electrical equipment—Every maintenance inspection.

3. Check operation of all lamps, direction indicators and horns.

   See Data Section for replacement bulb and units.

Seats, safety harness and rear view mirror—Every 5.000 km (3,000 miles) or 3 months.

4. Check all seat fixings for security and examine condition of safety harness. Safety harness which have been used in an accident or are frayed or cut must be replaced.

   Check rear view mirror(s) for security and examine mirror face for signs of cracking or crazing.

Door locks, bonnet release and window controls—At free service 1.500 km (1,000 miles) and thereafter every 10.000 km (6,000 miles) or six months.

5. Check operation of door locks, bonnet release control and window controls, rectify any faults as necessary.

   Apply a few spots of oil as required.

Headlamp beam setting—Every maintenance inspection. Fig.1.

This operation requires special equipment and should be carried out by your local Distributor or Dealer.

In an emergency each headlamp unit can be adjusted by means of:

1. **The headlamp horizontal adjusting screw.**
2. **The headlamp vertical adjusting screw.**

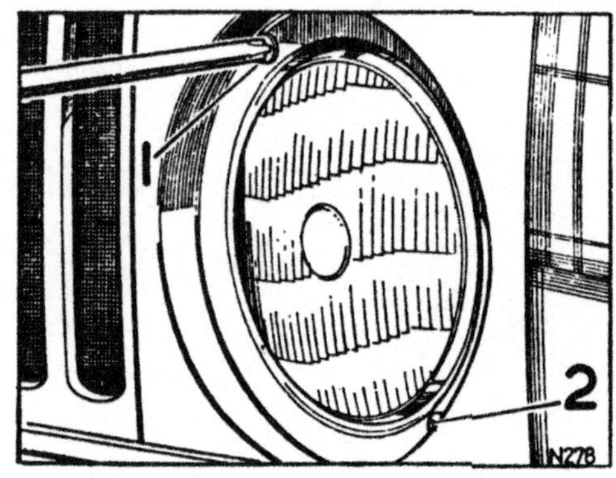

Fig. 1. Headlight unit adjustments

Headlamp wiper blades (where fitted)—Check, if necessary replace, every 5.000 km (3,000 miles) or 3 months.

To replace wiper blades:

1. Using finger pressure to prevent the blade from turning, release the centre screw.
2. Ease the centre frame away from the headlamp and withdraw the wiper arm and blades.
3. To fit new blades reverse removal procedure.

Windscreen and tailgate wiper blades—Check, if necessary replace, every 5.000 km (3,000 miles) or 3 months. Fig. 2.

To replace wiper blades:

1. Lift the wiper arm forwards, away from the windscreen
2. Twist the wiper fixing bracket in the direction arrowed and disengage it from the wiper arm.
3. To fit a new blade locate its fixing bracket over the end of the wiper arm and push on until the retaining dowel is engaged.

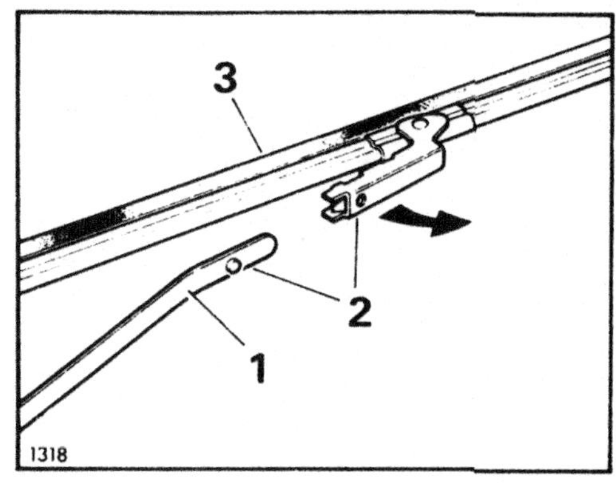

Fig. 2. Wiper blade replacement

Road wheels—Every maintenance inspection.

1. With the type of tyre used on Range Rover models, it is not considered advantageous to change the wheel positions; this in fact can give unpleasant handling characteristics when carried out, particularly if there is considerable difference between the wear pattern of one tyre and another.
2. Remove road wheels, wash and examine for possible damage.
3. For wheel removal, see page 107.

Before road wheel replacement, carry out the following operations.

Front brake pads—Every 5.000 km (3,000 miles) or 3 months. Fig. 3.

Hydraulic disc brakes are fitted at the front and the correct brake adjustment is automatically maintained; no provision is therefore made for adjustment.

1. Check the thickness of the front brake pads and renew if the minimum is less than 3,0 mm (0.125 in.).
2. Check for oil contamination on brake pads and discs, also check condition of brake discs for wear and/or corrosion.
3. If replacement or rectification is necessary, this should be carried out by your local Rover Distributor or Dealer.

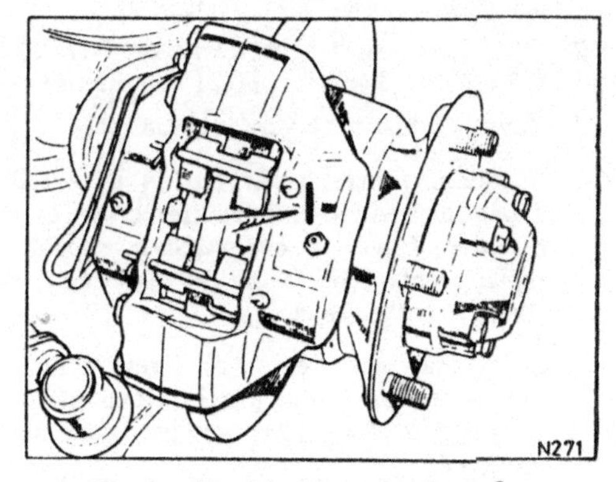

Fig. 3. Checking front brake pads

Rear brake pads—Every 5.000 km (3,000 miles) or 3 months. Fig. 4.

1. Hydraulic disc brakes are fitted at the rear and the correct brake adjustment is automatically maintained; no provision is therefore made for adjustment.
2. Check the thickness of the rear brake pads and renew if the minimum is less than 1,5 mm (0.062 in.).
3. Check for oil contamination on brake pads and discs and check condition of discs for wear and/or corrosion.
   Also check brake anti-squeal shims for corrosion.
4. If replacements or rectification is necessary, this should be carried out by your local Rover Distributor or Dealer.

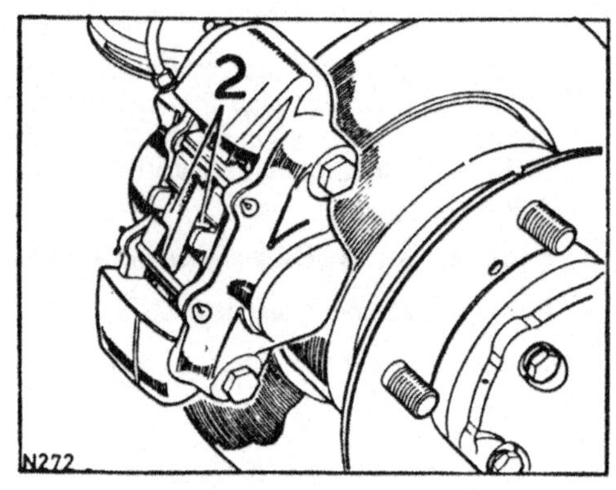

Fig. 4. Checking rear brake pads

Tyre pressures

1. These should be checked at least every month for normal road use and at least weekly, preferably daily, if the vehicle is used off the road. (See General Data, Section 9).
2. Whenever possible check with the tyres cold, as the pressure is about 0,2 kg/cm² (3 lb/sq in.) 0,21 bars higher at running temperature.
3. Always replace the valve caps, as they form a positive seal.
4. **Warning: Range Rover models have radial-ply tyres as standard fitting, and whenever replacements are required radial-ply tyres must be fitted. Under no circumstances should cross-ply tyres be used as replacements. Always use the same make and type of radial-ply tyre throughout the vehicle.**
5. Check that pressures on all tyres, including the spare, are correct. Any unusual pressure loss in excess of 0,05 to 0,20 kg/cm² (1 to 3 lb/sq in.) 0,07 to 0,21 bars per week should be investigated and corrected.

When high-speed touring the pressures should be checked much more frequently, even to the extent of a daily check. If front tyre tread is uneven, check wheel alignment.

Tyre wear

6. Most tyres fitted to Range Rovers as original equipment include wear indicators in their tread pattern. When the tread has worn to a remaining depth of 1,6 mm (1/16") the indicators appear at the surface as bars which connect the tread pattern across the full width of the tyre, as in the Goodyear tyre section illustrated. When the indicators appear in two or more adjacent grooves, at three locations around the tyre, a new tyre should be fitted. If the tyres do not have wear indicators, the tread should be measured at every maintenance inspection and when the tread has worn to a remaining depth of 1,6 mm (1/16"), new tyres should be fitted. Do not continue to use tyres that have worn to the recommended limit or the safety of the vehicle could be affected and legal regulations governing tread depth may be broken.

7. Check that there are no lumps or bulges in the tyres or exposure of the ply or cord structure.
   Clean off any oil or grease, using white spirit sparingly.

8. At the same time remove embedded flints, etc. from the treads with the aid of a penknife or similar tool, and check that the tyres have no 'breaks' in the fabric or cuts to sidewalls, etc.

9. It is advisable to run-in new tyres by driving at reasonable speeds for the first 400 km (250 miles) or so before driving at high speeds.

10. Wheel and tyre units are accurately balanced on initial assembly with the aid of clip-on weights secured to the wheel rims.

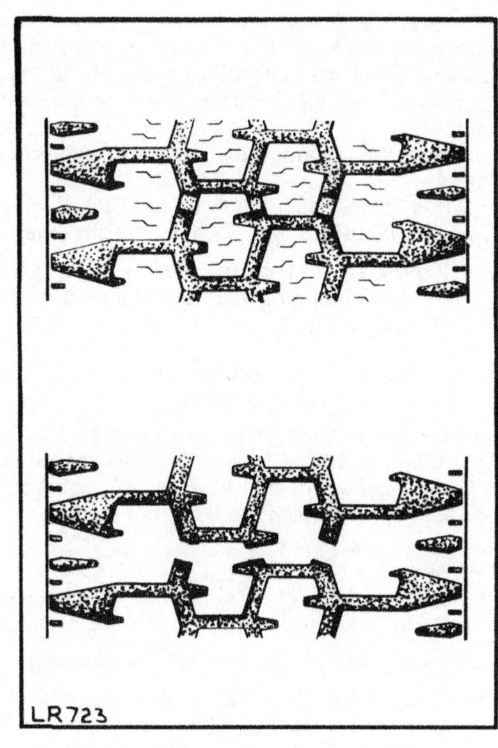

Fig. 4a. New and worm tread showing typical indicator bars

Warning: The Range Rover has permanent four-wheel drive. It is most important therefore that before any attempt is made to dynamically balance the wheels in position on the vehicle, the gearbox differential must be locked and the propeller shaft to the stationary wheels must be removed.

Failure to comply with these instructions could result in personal injury.

Wheel balance should always be checked whenever new tyres are fitted to ensure that the dynamic balance of the wheel and tyre are correct.

11. Replace road wheels in original position and finally check tightness of all road wheel nuts to a torque figure of between 10,0 and 11,7 kgf. m. (75 and 85 lbf. ft.).

Wheel alignment—At free service 1.500 km (1,000 miles) and thereafter every 10.000 km (6,000 miles) or 6 months. Fig. 5.

Special equipment is required to check wheel alignment and this work should be carried out by your local Rover Distributor or Dealer.

For those owners who have suitable equipment, the alignment should be 1,2 to 2,4 mm (0.046 to 0.093 in.) toe-out.

To adjust

1. Set the vehicle on level ground, with the road wheels in the straight-ahead position, and push it forward a short distance.
2. Slacken the clamps securing the adjusting shaft to the track rod.
3. Turn the adjusting shaft to decrease or increase the effective length of the track rod as necessary, until the toe-out is correct.
4. Re-tighten the clamps.
5. Push the vehicle rearwards, turning the steering wheel from side to side to settle the ball joints. Then with the road wheels in the straight-ahead position, push the vehicle forward a short distance.
6. Recheck the toe-out. If necessary carry out further adjustment.

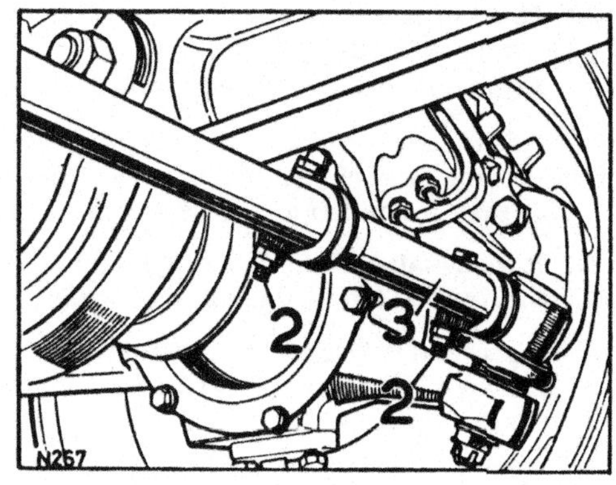

Fig. 5. Wheel alignment

Range Rover maintenance location points for:

7 Engine sump drain plug
8 Steering ball joints
9 Front differential oil level, filler and drain plugs
10 Swivel pin housing oil level/filler and drain plugs
11 Flywheel housing drain plug
12 Gearbox drain plug and filter—plus oil level/filler plug
13 Transfer box oil level and filler plug
14 Transfer box drain plug
15 Transmission brake adjustment
16 Fuel pump filter cleaning
17 Propeller shaft lubrication
18 Rear differential oil filler, level and drain plugs
19 Engine oil filter
20 Axle case breathers

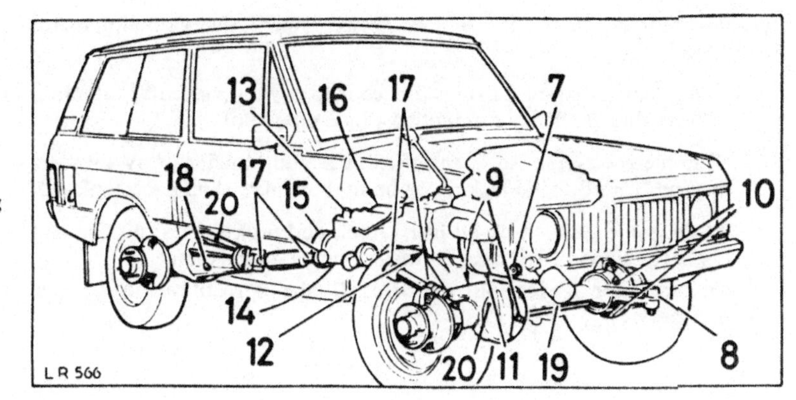

Every maintenance inspection

Examine underbody components for oil leaks; rectify as necessary.

Engine oil changes—At free service 1.500 km (1,000 miles) and thereafter every 10.000 km (6,000 miles) or 6 months. Fig. 6.

To change the engine oil:

1. Run the engine to warm up the oil; switch off the ignition.
2. Remove the drain plug in the bottom of the sump at left-hand side. Allow oil to drain away completely and replace the plug.
3. Replenish the sump through the screw-on filler cap marked 'engine oil' on the front right-hand rocker cover, using fresh oil of the correct grade. (See General Data, Section 9 – Capacities).

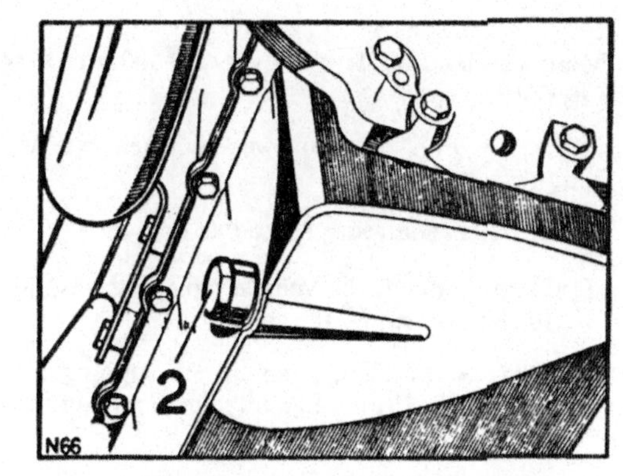

Fig. 6. Engine sump drain plug

Exhaust system, fuel, clutch and brake pipes—Every maintenance inspection.

1. Check exhaust system fixings for security paying particular attention to heat shields, flexible mountings, and pipe joints.
2. Examine the system for signs of leakage and blowing. Any silencers or pipes found to be leaking or badly corroded should be replaced.
3. At the same time check all fuel, clutch and brake pipes, unions and hoses for signs of leakage, corrosion, chafing or damage.

   Contact your Rover Distributor or Dealer if rectification work is necessary.

Main gearbox oil level—Every 10.000 km (6,000 miles) or 6 months. Fig. 7.

Check oil level daily or weekly when operating under severe wading conditions.

If oil is required, proceed as follows:

1. From beneath the vehicle remove the oil level/filler plug and top up to the bottom of the hole.
2. Replace the plug. If significant topping up is required check for oil leaks at the drain plug, all joint faces and through the drain hole in the bell housing.

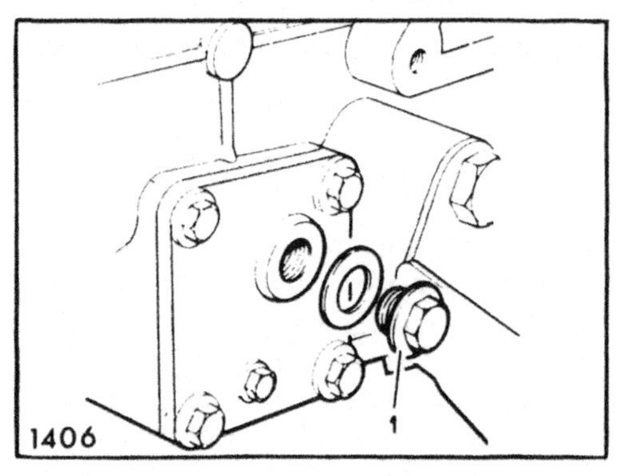

Fig. 7. Gearbox oil level/filler plug

Steering ball joints—Every maintenance inspection. Fig. 8.

Check rubber boots daily when operating under arduous conditions.

1. The steering joints have been designed to retain the initial filling of grease for the normal life of the ball joints; however, this applies only if the rubber boot remains in the correct position.

2. Check to ensure that the rubber boots have not become dislodged or damaged, and check for wear in the joint.

   This can be done by moving the ball joint vigorously up and down. Should there be any appreciable free movement the complete joint must be replaced.

Fig. 8. Ball joints

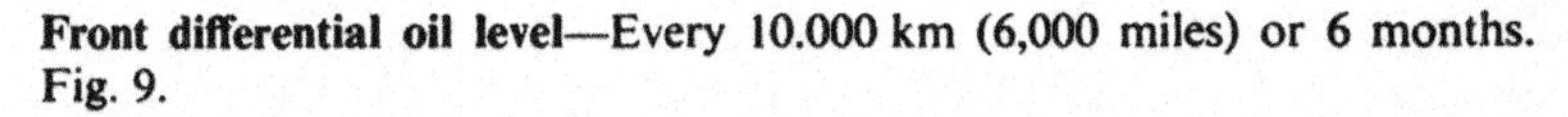

Front differential oil level—Every 10.000 km (6,000 miles) or 6 months. Fig. 9.

1. Remove filler/level plug, check oil level and top-up, if necessary, to the bottom of the filler/level plug hole.

2. If significant topping-up is required, check for oil leaks at plugs, joint faces, and oil seals adjacent to axle shaft flanges and propeller shaft driving flange.

Front differential oil changes—At free service 1.500 km (1,000 miles) and thereafter every 40.000 km (24,000 miles) or 24 months. Fig. 9.

3. Drain and refill monthly under severe wading conditions.

   To change the differential oil, proceed as follows:

4. Immediately after a run when the oil is warm drain off the oil by removing the drain plug.

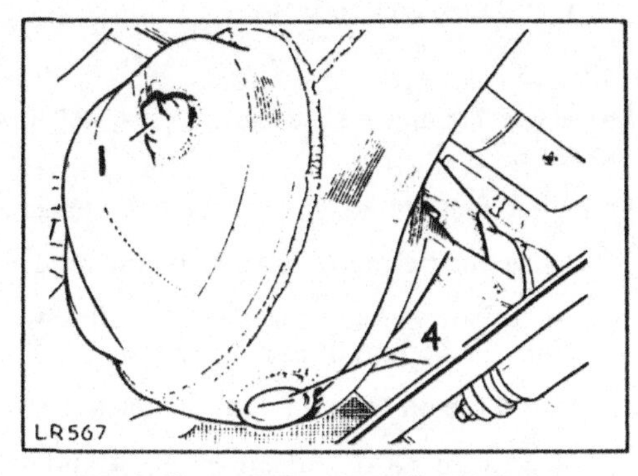

Fig. 9. Front differential oil filler/level and drain plug

5. Replace the drain plug and refill with oil of the correct grade. The capacity is approximately 1,7 litres (3 Imperial pints) 3.5 US pints.

   Important: Do not overfill otherwise damage to the seals may occur

The drain plug has a slotted head and can be removed with the aid of a single-ended spanner.

Swivel pin housing oil level—Every 10.000 km (6,000 miles) or 6 months. Fig. 10.

1. The front wheel drive universal joints and swivel pins receive their lubrication from the swivel pin housing.
2. Check oil level by removing the $\frac{1}{4}$ in. AF square-headed plug at the front of the swivel pin housing; oil should be level with the bottom of the hole.
3. Top up if necessary through the filler plug hole.

   If significant topping up is required, check for oil leaks at plugs, joint faces, and oil seals.

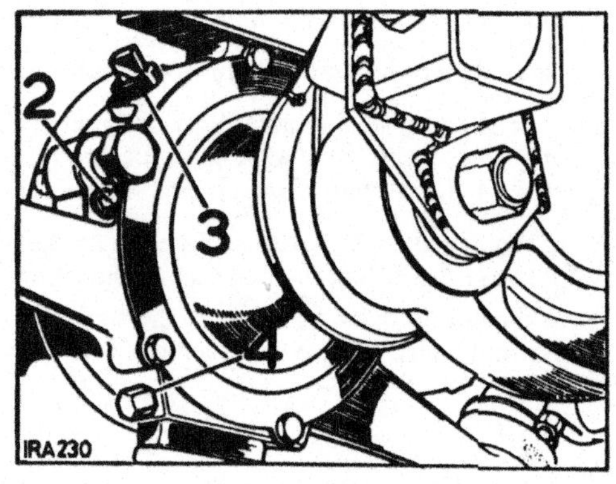

Fig. 10. Swivel pin housing oil/level and drain plugs

Swivel pin housing oil changes—Every 40.000 km (24,000 miles) or 24 months. Fig. 10.

Drain and refill monthly when operating under severe wading conditions.

To change the swivel pin housing oil, proceed as follows:

4. Immediately after a run, when the oil is warm, remove the drain plug from the bottom of each housing.
5. Allow the oil to drain away completely and replace drain plugs.
6. Refill with oil of the correct grade through the oil filler/level plug hole. The capacity of each housing is approximately 0,26 litres (0.5 Imperial pint), 0.5 US pint.

Flywheel housing drain plug—Every 5.000 km (3,000 miles) or 3 months. Fig. 11.

When in use for wading:

1. The flywheel housing can be completely sealed to exclude mud and water under severe wading conditions, by means of a plug fitted in the bottom of the housing.
2. The plug is screwed into the housing adjacent to the drain hole, and should only be fitted when the vehicle is expected to do wading or very muddy work.

   When the plug is in use it must be removed periodically and all oil allowed to drain off before the plug is replaced.

Fig. 11. Flywheel housing drain plug

Main gearbox oil changes—At free service 1.500 km (1,000 miles) and thereafter every 40.000 km (24,000 miles) or 24 months. Fig. 12.

Drain and refill monthly when operating under severe wading conditions.

To change the gearbox oil proceed as follows:

1. Immediately after a run when the oil is warm, drain off the oil by removing the drain plug and washer from the bottom of the gearbox casing.
2. Remove the oil filter.
3. Wash the filter in clean fuel; allow to dry and replace.
4. Refit drain plug and washer and refill gearbox through the oil level/filler plug, with the correct grade of oil, to the bottom of the oil level/filler hole. See page 54.

   The capacity is 2,6 litres (4.5 Imperial pints) 5.5 US pints.

Important: Do not overfill, otherwise leakage may occur.

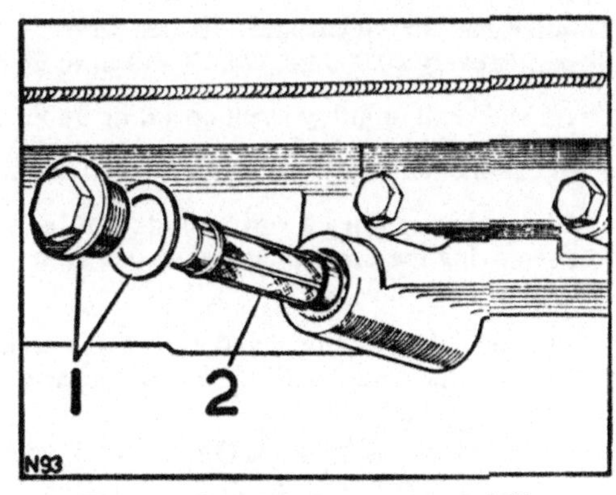

Fig. 12. Gearbox drain plug and filter

Transfer box oil level—Every 10.000 km (6,000 miles) or 6 months. Fig. 13.

Check oil level daily or weekly when operating under severe wading conditions.

1. To check oil level: remove the oil level plug, located on the rear of the transfer box casing; oil should be level with the bottom of the hole.
2. To top up: remove the round rubber blanking plug from the gearbox cover.
3. Remove the oil filler plug from the transfer box, and top up as necessary. If significant topping up is required, check for oil leaks at drain and filler plugs.

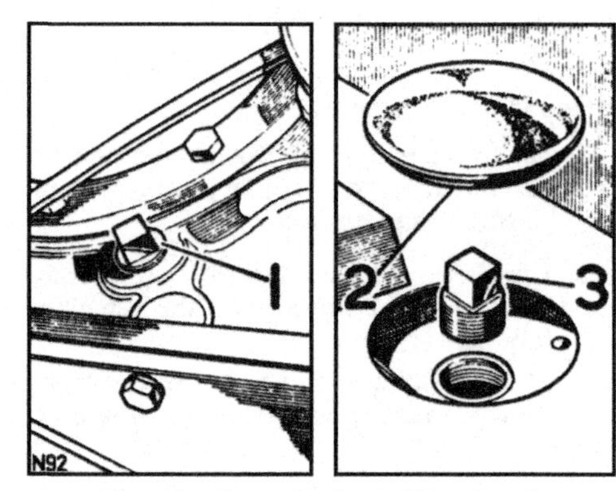

Fig. 13. Transfer box oil level and filler plug

Transfer gearbox oil changes—At free service 1.500 km (1,000 miles) and thereafter every 40.000 km (24,000 miles) or 24 months. Fig. 14.

Drain and refill monthly when operating under severe wading conditions.

To change the transfer box oil, proceed as follows:

1. Immediately after a run when the oil is warm, drain off the oil by removing the drain plug and washer from the bottom of the transfer box.
2. Replace the drain plug and washer and refill the transfer box through the oil filler plug, with the correct grade of oil, to the bottom of the oil level plug hole.

   The capacity is 3,1 litres (5.5 Imperial pints) 6.5 US pints.

Important: Do not overfill otherwise leakage may occur.

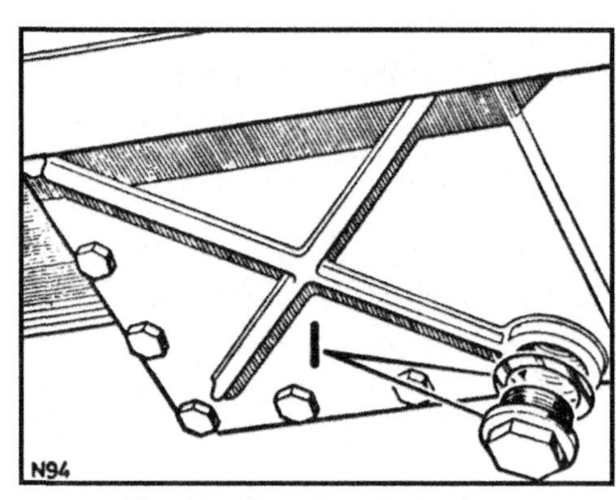

Fig. 14. Transbox drain plug

Transmission fixings—At free service 1.500 km (1,000 miles) only.

Check security of transmission fixings; rectify as necessary.

Handbrake linkage—Every 10.000 km (6,000 miles) or 6 months.

The handbrake operates a mechanical brake unit mounted on the output shaft from the transfer box. Lubricate the handbrake linkage and check for worn parts. Take care not to contaminate the handbrake linings with oil.

Transmission brake adjustment—Fig. 15.

If the handbrake movement is excessive adjust as follows:

1. Set the vehicle on level ground.
2. Release the handbrake fully.
3. From beneath the vehicle, remove the rubber blanking plugs from the brake drum.
4. Move the vehicle either forwards or backwards until the adjuster can be seen through one of the apertures.
5. With a screwdriver turn the adjuster wheel until the brake shoes come into contact with the brake drum.
6. Turn the adjuster back two 'clicks' and replace blanking plugs.
7. Check that the handbrake operates correctly and holds the vehicle.

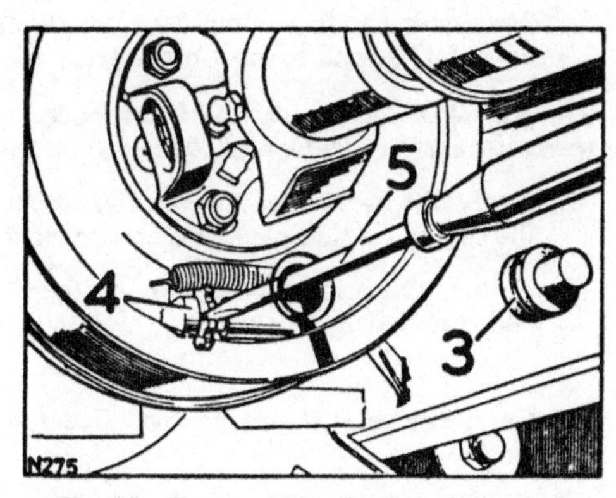

Fig. 15. Transmission brake adjustment

Electric fuel pump filter—Clean every 80.000 km (48,000 miles) or 48 months. Fig. 16.

The electric fuel pump is located on the heelboard beneath the rear seat, left-hand side.

To remove the filter for cleaning proceed as follows:

1. From beneath the vehicle disconnect the fuel inlet pipe from the pump and blank the end of the pipe by suitable means to prevent fuel draining from the tank.
2. Release the end cover from the bayonet fixing using a 0.625 in. AF spanner.
3. Withdraw the filter and clean by using a compressed air jet from the inside of the filter.
4. Remove the magnet from the end cover and clean. Replace the magnet in the centre of the end cover.
5. Reassemble the fuel pump and refit the fuel inlet pipe. Use a new gasket for the end cover if necessary.

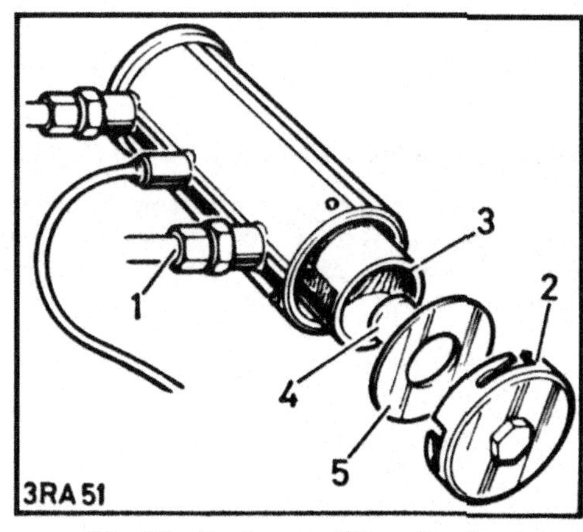

Fig. 16. Fuel pump filter cleaning

Propeller shaft lubrication—At free service 1.500 km (1,000 miles) and thereafter every 10.000 km (6,000 miles) or 6 months. Fig. 17.

1. Apply one of the recommended greases at the lubrication nipple on the sliding portion of the rear propeller shaft.
2. To the lubrication nipples fitted to the universal joints of both front and rear shafts.

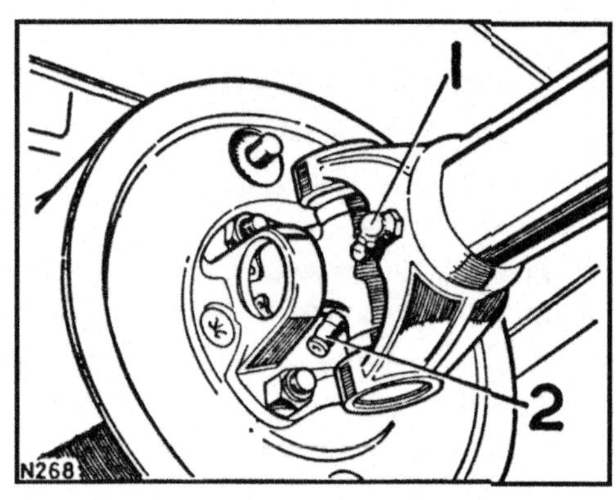

Fig. 17. Propeller shaft lubrication

Front propeller shaft sliding portion—Every 40.000 km (24,000 miles) or 24 months.

Lubricate the sliding spline on the front propeller shaft, with one of the recommended greases, as follows:

3. **Disconnect one end of the propeller shaft.**
4. **Remove plug in sliding spline and fit a suitable grease nipple.**
5. **Important: Compress propeller shaft at sliding joint to avoid overfilling, then apply grease.**
6. **Replace grease nipple with plug and reconnect propeller shaft.**

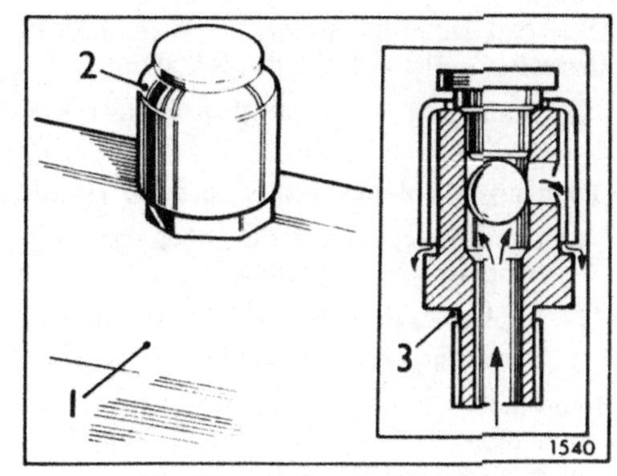

Fig. 17a. Axle case breathers

Axle case breathers—Every 20,000 km or 12 months.
Clean the axle case breathers, one in each axle case.

1. Clean off the axle breathers and the surrounding surfaces of the axle cases taking care to remove any gritty foreign matter.
2. Unscrew the axle breathers from their tapered threads in the axle tubes and soak in petrol or a suitable cleaning solvent for several minutes and clean with a soft brush.
3. Shake each breather to ensure the ball valve is free. If it is not the breather valve must be renewed.
4. Lubricate the balls lightly with engine oil before replacing the breathers.

Rear differential oil level—Every 10.000 km (6,000 miles) or 6 months. Fig. 18.

1. Check oil level and top-up if necessary to the bottom of the filler/level plug hole, located on the right-hand side of the differential casing (as illustrated).

   Note: On certain axles the filler/level plug is located in the rear casing of the differential, as on front axles (see Fig. 9, page 55).

2. If significant topping-up is required, check for oil leaks at plugs, joint faces, and oil seals adjacent to axle shaft flanges and propeller shaft driving flange.

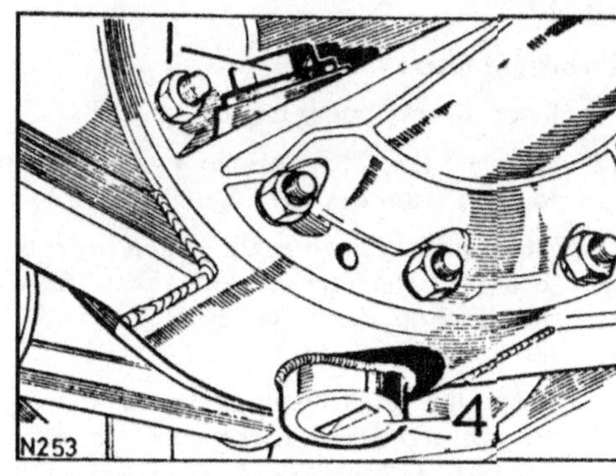

Fig. 18. Rear differential oil filler/level and drain plugs

Rear differential oil changes—At free service 1.500 km (1,000 miles) and thereafter every 40.000 km (24,000 miles) or 24 months. Fig. 18.

3. Drain and refill monthly when operating under severe wading conditions.

To change the differential oil, proceed as follows:

4. Immediately after a run, when the oil is warm, drain off the oil by removing the drain plug.
5. Replace drain plug and refill with oil of the correct grade.
   The capacity is approximately 1,7 litres (3 Imperial pints) 3.5 US pints.

Important: Do not overfill otherwise damage to seals may occur. The drain plug has a slotted head and can be removed with the aid of a single-ended spanner.

Engine oil filter replacement—Every 10.000 km (6,000 miles) or 6 months. Fig. 19.

To change filter:

1. Place oil tray under engine.
2. Unscrew the filter anti-clockwise and discard. It may be necessary to use a strap spanner or similar tool to release the filter.
3. Smear a little clean engine oil on the rubber washer of the new filter, then screw the filter on clockwise until the rubber sealing ring touches the oil pump cover face, then tighten a further half turn by hand only. Do not overtighten.
4. Refill with oil of the correct grade through the screw-on filler cap on the right-hand front rocker cover; the capacity is 5,5 litres (10 Imperial pints), 12 US pints. This includes 0,5 litres (1 Imperial pint), 1.2 US pints, for the filter.

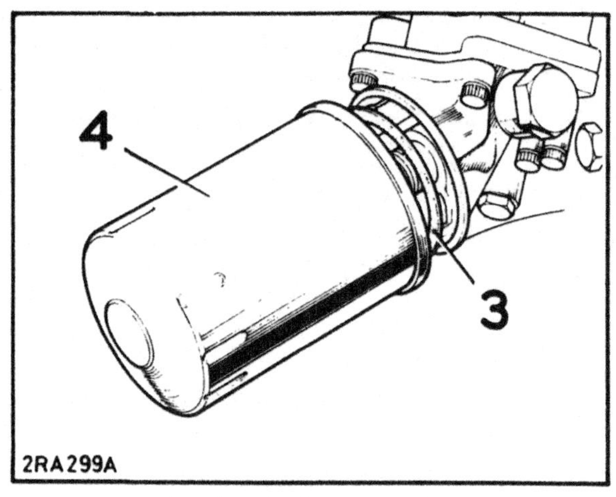

Fig. 19. Oil filter for engine

Range Rover maintenance location points for:

20 Coolant expansion tank filler cap
21 Radiator drain tap
22 Cylinder block drain plug
23 Radiator filler plug
24 Clutch fluid reservoir
25 Brake fluid reservoir
26 Steering box lubrication
27 Battery
28 Alternator

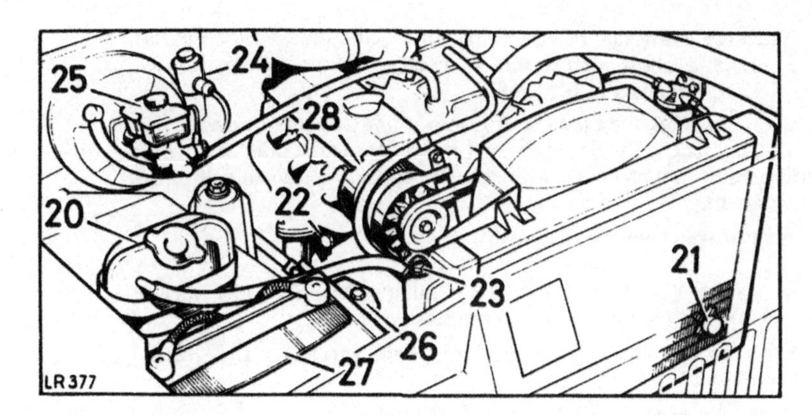

Radiator coolant level—Every 1.000 km (750 miles) and at every maintenance inspection. Fig. 20.

1. **To prevent corrosion of the aluminium alloy engine parts it is imperative that the cooling system is filled with a solution of clean water and the correct type of anti-freeze, winter and summer, or water and inhibitor if frost precautions are not required. Never fill or top-up with water only, always add an inhibitor (Marstons SQ36) if anti-freeze is not used.**

CAUTION: **Do not use salt water even with an inhibitor, otherwise corrosion will occur. In certain territories where the only available natural water supply has some salt content use only rain or distilled water.**

2. The expansion tank filler cap is under the bonnet.
3. With a cold engine, the correct coolant level should be up to the 'Water Level' plate situated inside the expansion tank below the filler neck.

WARNING: **Do not remove the filler cap when the engine is hot because the cooling system is pressurised and personal scalding could result.**

Fig. 20. Expansion tank filler cap and 'Water Level' plate

4. When removing the filler cap, first turn it anti-clockwise a quarter of a turn and allow all pressure to escape, before turning further in the same direction to lift it off.
5. When replacing the filler cap it is important that it is tightened down fully, not just to the first stop. Failure to tighten the filler cap properly may result in water loss, with possible damage to the engine through overheating.

   Use soft water whenever possible.

Cooling system—At free service 1.500 km (1,000 miles) and thereafter every 10.000 km (6,000 miles) or 6 months.

1. Examine the cooling and heater system for leaks and rectify as necessary.
2. Renew hoses every 80.000 km (48,000 miles).

Frost precautions and engine protection—Figs. 21, 22 and 23.
During both the winter and summer months special anti-freeze mixture is used in Range Rover models to prevent corrosion of the aluminium alloy engine parts. **It is most important, therefore, if the cooling system is drained or topped up at any time, to use a solution of water and anti-freeze during winter and summer, or water and inhibitor during the summer only.**

Recommended solutions are:
Anti-freeze—Unipart Universal Antifreeze.
Inhibitor—Marston Lubricants SQ36. Coolant inhibitor concentrate.
Use one part of anti-freeze to two parts of water.
Use 50 cc of inhibitor per litre of water (8 fluid ounces of inhibitor per gallon of water). Inhibitor solution should be drained and flushed out and new inhibitor solution introduced every two years or sooner where the purity of the water is questionable.

Anti-freeze can remain in the cooling system and will provide adequate protection for two years provided that the specific gravity of the coolant is checked before the onset of the second winter and topped up with new anti-freeze as required.

After the second winter the system should be drained and thoroughly flushed by using a hose inserted in the radiator filler orifice. Before adding new anti-freeze examine all joints and renew defective hoses to make sure that the system is leakproof.

At the lower limit of protection, a mixture of water and anti-freeze will reach a 'mushy' state with a viscosity that can impair circulation and immobilize or damage the water pump. Therefore, consult the following chart for concentration of anti-freeze required to protect the system at temperatures likely to be encountered.

	ANTI-FREEZE CONCENTRATION	25%	30%	35%	50%
	SPECIFIC GRAVITY OF COOLANT AT 15.50°C (60°F)	1.039	1.048	1.054	1.076
DEGREE OF PROTECTION	**Complete** Car may be driven away immediately from cold	−12°C 10°F	−16°C 3°F	−20°C − 4°F	−36°C −33°F
DEGREE OF PROTECTION	**Safe Limit** Coolant in a mushy state. Engine may be started and driven away after short warm-up period	−18°C 0°F	−22°C − 8°F	−28°C −18°F	−42°C −42°F
DEGREE OF PROTECTION	**Lower Protection** Prevents frost damage to cylinder head, block and radiator. Thaw out before starting the engine	−26°C −14°F	−32°C −25°F	−37°C −35°F	−47°C −53°F

To change the solution proceed as follows:

1. Ensure that the cooling system is leak-proof; anti-freeze solutions are far more searching at joints than water.
2. Drain and flush the system, radiator drain plug located on bottom left-hand side.

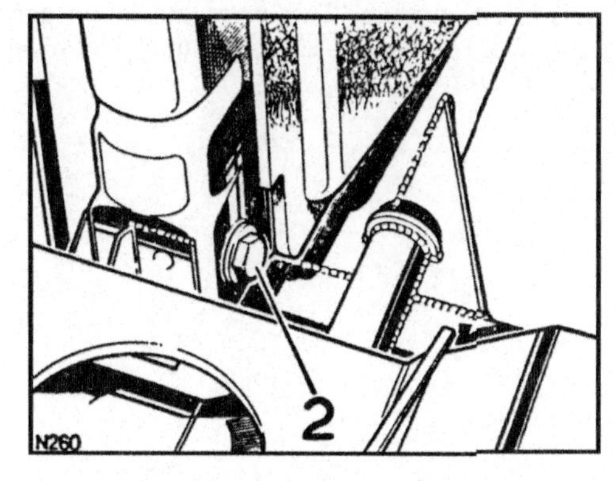

Fig. 21. Radiator drain plug

3. Drain plugs are on the right and left sides of the cylinder block.

4. Remove radiator filler plug and washer located on top right-hand side.

5. Replace drain plugs and pour in approximately 4,5 litres (one gallon) of water. Add the recommended quantity of anti-freeze or inhibitor if frost protection is not required.

6. Top up the radiator with water, refit the radiator filler plug and washer securely.

7. Add water to the expansion tank, up to the 'Water Level' plate, and replace cap.

8. Run the engine until normal operating temperature is attained, that is, thermostat open. Allow the engine to cool, then check the coolant level and top up if necessary.

Range Rover models have the cooling system filled with 33⅓% of anti-freeze mixture. This gives protection against frost down to minus 32°C (25°F). Vehicles so filled can be identified by the green label tied to the radiator.

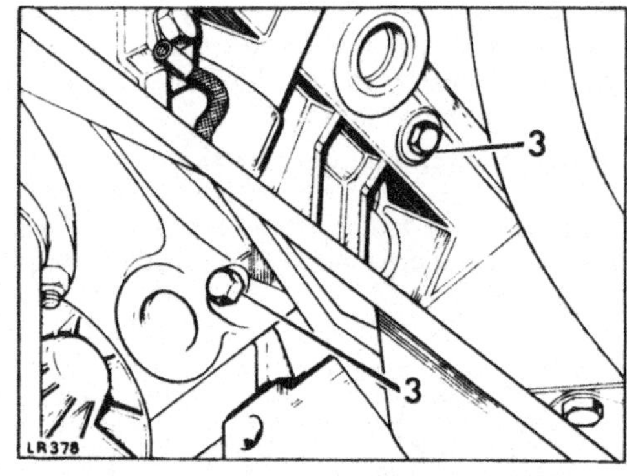

Fig. 22. Cylinder block drain plugs

Fig. 23. Radiator filler plug

Every maintenance inspection
Check for oil leaks in engine compartment; rectify as necessary.

Clutch fluid reservoir—Every maintenance inspection. Fig. 24.

1. Check the fluid level in the reservoir, mounted on the bulkhead adjacent to the brake servo.
2. Remove the cap; top up if necessary to bottom of filler neck.
   Use Unipart Universal Brake (and Clutch) Fluid or other brake fluids having a minimum boiling point of 260°C (500°F) and complying with FMVSS 116 DOT3 or SAE J1703 specification.
   If significant topping up is required, check for leaks at master cylinder, slave cylinder, and connecting pipes.

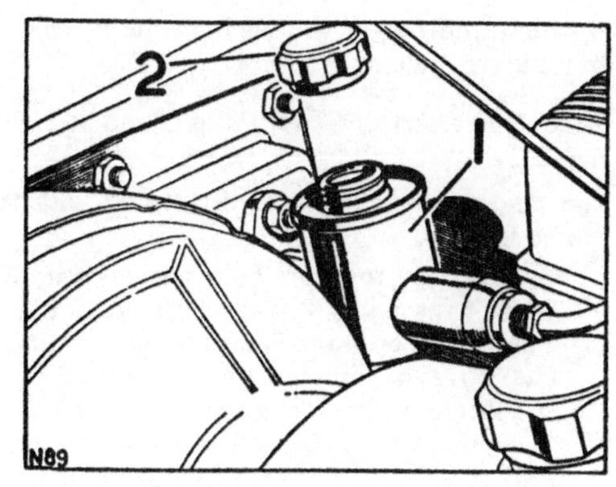

Fig. 24. Clutch fluid reservoir

Braking system

The primary and secondary dual line braking system fitted to the Range Rover is designed to function should there be a failure of one or more of its component parts.

For example, in the event of failure to a component part of the primary system, the secondary braking system will continue to operate. Conversely, in the event of a failure to a component on the secondary brakes, braking will still be available to the primary system.

The system consists of a brake servo tandem reservoir unit fitted with a master cylinder, located on the right-hand bulkhead. This assembly feeds a reaction valve attached to the right-hand wing valance, and thence to each wheel cylinder.

The red warning light on the instrument panel marked 'BRAKE' is most important, and is arranged to warn you of a fluid leakage from either the primary or secondary braking system. Should there be a leakage the warning light will come on when the foot brake is applied, and go out when pedal pressure is released.

The warning light will also operate if a loss of vacuum occurs in the brake servo system.

Brake fluid reservoir—Every month and at every maintenance inspection Fig. 25.

The tandem brake reservoir is integral with the servo unit and master cylinder.

1. Remove cap to check fluid level; top up if necessary until the fluid reaches the bottom of the filler neck. Use Unipart Universal Brake Fluid or other brake fluids having a minimum boiling point of 260°C (500°F) and complying with FMVSS 116 DOT3 or SAE J1703 specification, from sealed tins.
2. If significant topping up is required check master cylinder, brake disc cylinders and brake pipes and connections for leakage; any leakage must be rectified immediately.

Caution. When topping up the reservoir, care should be taken to ensure that brake fluid does not come into contact with any paintwork on the vehicle.

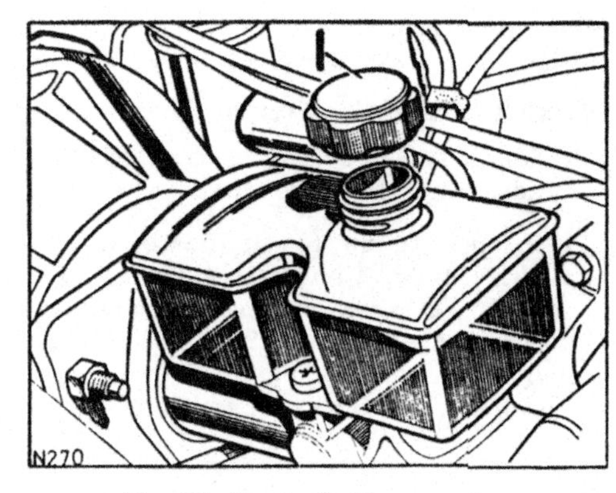

Fig. 25. Brake fluid reservoir

Steering box lubrication (manual steering)—At free service 1.500 km (1,000 miles) and thereafter every 20.000 km (12,000 miles) or 12 months. Fig. 26.

1. Check oil level and top-up if necessary to 12,7 mm (0.5 in.) below the filler plug hole on the top of the cover plate. Do not overfill otherwise leakage may occur.
2. If significant topping up is required check for oil leaks at joint faces and rocker shaft oil seal.

Steering unit—Every maintenance inspection.

Check condition of steering unit fixings for security, rectify as necessary.

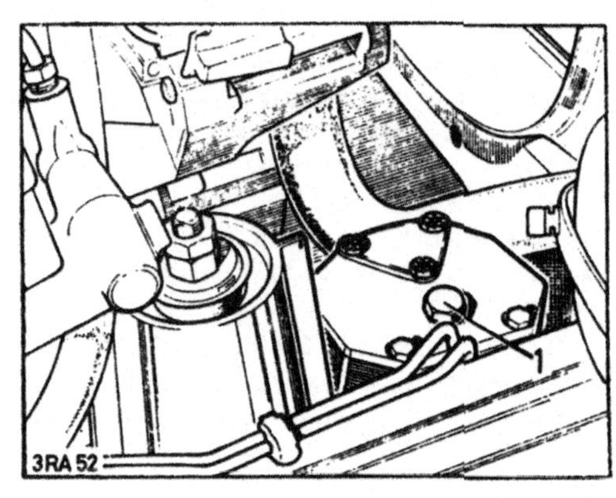

Fig. 26. Manual steering box lubrication

Battery acid level—Every month and at every maintenance inspection. Fig. 27.

The specific gravity of the electrolyte should be checked at every maintenance inspection.

Readings should be:

Temperate climates below 26.5°C (80°F) as commissioned for service, fully charged 1.270 to 1.290 specific gravity.

As expected during normal service three-quarter charged 1.230 to 1.250 specific gravity.

If the specific gravity should read between 1.190 to 1.210, half-charged, the battery must be bench charged and the electrical equipment on the vehicle should be checked.

Tropical climate, above 26.5°C (80°F) as commissioned for service, fully charged 1.210 to 1.230 specific gravity.

As expected during normal service three-quarter charged 1.170 to 1.190 specific gravity.

If the specific gravity should read between 1.130 to 1.150, half-charged, the battery must be bench charged and the electrical equipment on the vehicle should be checked.

The battery is located under the bonnet.

Check acid level as follows:

1. Remove the battery lid.
2. If necessary add sufficient distilled water to raise the level to the top of the separators; do NOT overfill.
3. Avoid the use of naked lights when examining the cells.
4. In hot climates it will be necessary to top up the battery at more frequent intervals.
5. In very cold weather it is essential that the vehicle is used immediately after topping up to ensure that the distilled water is thoroughly mixed with the electrolyte. Neglect of this precaution may result in the distilled water freezing and causing damage to the battery.

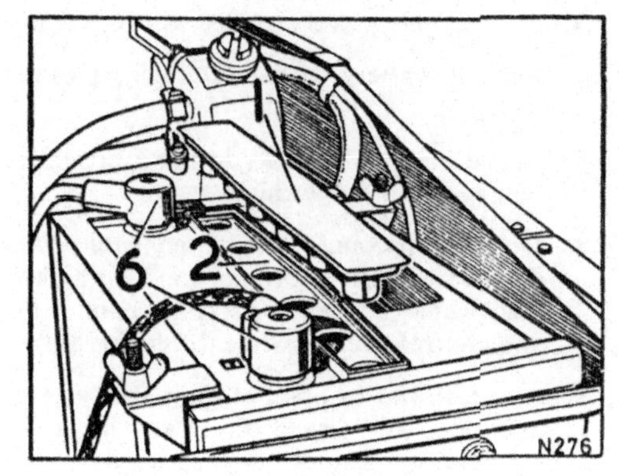

Fig. 27. Battery

Battery terminals—Every 10.000 km (6,000 miles) or 6 months. Fig. 27.

6. Remove battery terminals, clean, grease with petroleum jelly and refit.
7. Replace terminal screw; do not overtighten. Do not use the screw for pulling down the terminal.
8. Do NOT disconnect the battery cables while the engine is running or damage to alternator semiconductor devices may occur. It is also inadvisable to break or make any connection in the alternator charging and control circuits while the engine is running.
9. It is essential to observe the polarity of connections to the battery, alternator and regulator, as any incorrect connections made when reconnecting cables may cause irreparable damage to the semi-conductor devices.

Alternator—Every 20.000 km (12,000 miles) or 12 months. Fig. 28.

1. The alternator is a sealed unit, and requires no lubrication or maintenance.
2. Check and ensure that any dirt or oil which may have collected around the apertures in the slip-ring end bracket and moulded cover is wiped clear.

   Note. Alternator charging circuit
   The ignition warning light is connected in series with the alternator field circuit. Bulb failure would prevent the alternator charging, except at very high engine speeds, therefore the bulb should be checked before suspecting an alternator fault.

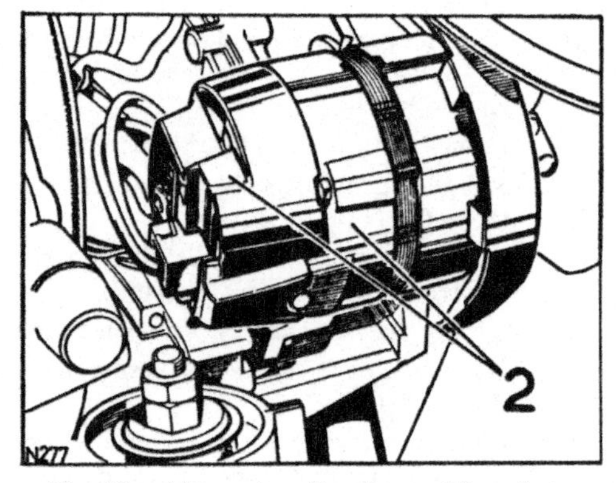

Fig. 28. Alternator slip-ring end bracket and moulding cover

Range Rover maintenance location points for:

29 Spark plug removal
30 Spark plug cleaning
31 } Distributor
32 } contact points
33 Distributor maintenance
34 Sequence of distributor leads
35 Layout of high tension leads
36 Dwell angle adjustment
37 Setting ignition timing

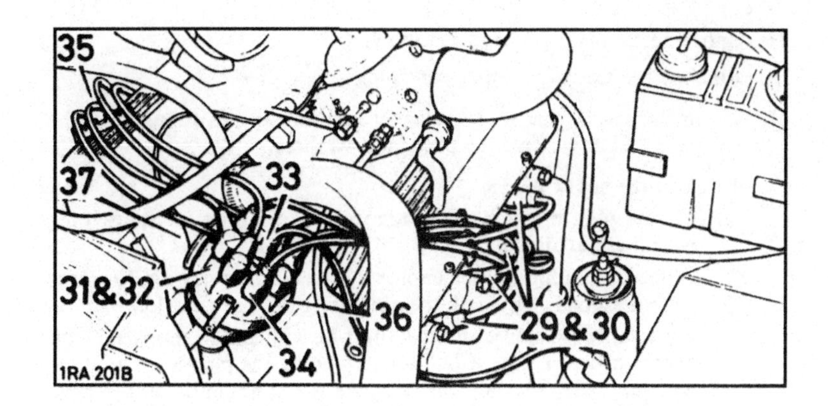

Spark plugs—Check every 10.000 km (6,000 miles) or 6 months. Replace every 20.000 km (12,000 miles) or 12 months. Figs. 29 and 30.

1. **Use the special spark plug spanner and tommy bar** supplied in the tool kit when removing or refitting spark plugs.
2. **Take great care when fitting spark plugs not to cross-thread the plug, otherwise costly damage to the cylinder head will result.**
3. Check or replace the spark plugs as applicable. If the plus are in good condition, clean and reset the electrode gaps to 0,80 mm (0.030 in.), at the same time file the end of the central electrode until bright metal can be seen.
4. It is important that only spark plugs specified in 'General Data', Section 9 are used for replacements.
5. Incorrect grades of plug may lead to piston over-heating and engine failure.

To remove spark plugs proceed as follows:

6. Remove the leads from the spark plugs.
7. Using the **special spark plug spanner and tommy bar** supplied in the vehicle tool kit, remove the plugs and washers.
8. To clean the spark plugs:
   (a) Fit the plug into a 14 mm adaptor of an approved spark plug cleaning machine.
   (b) Wobble the plug in the adaptor with a circular motion for **three or four seconds only** with the abrasive blast in operation.
   Important: Excessive abrasive blasting will lead to severe erosion of the insulator nose. Continue to wobble the plug in its adaptor with **air only**, blasting the plug for a minimum of **30 seconds;** this will remove abrasive grit from the plug cavity.
   (c) Wire-brush the plug threads; open the gap slightly, and vigorously file the electrode sparking surfaces using a point file. This operation is important to ensure correct plug operation by squaring the electrode sparking surfaces.
9. Set the electrode gap to the recommended clearance of 0,80 mm (0.030 in.).
10. Shows dirty plug.
11. Filing plug electrodes.
12. Clean plug set to correct gap.
13. Test the plugs in accordance with the plug cleaning machine manufacturer's recommendations.
14. If satisfactory the plugs may be replaced in the engine.
15. When pushing the leads on to the plugs, ensure that the shrouds are firmly seated on the plugs.

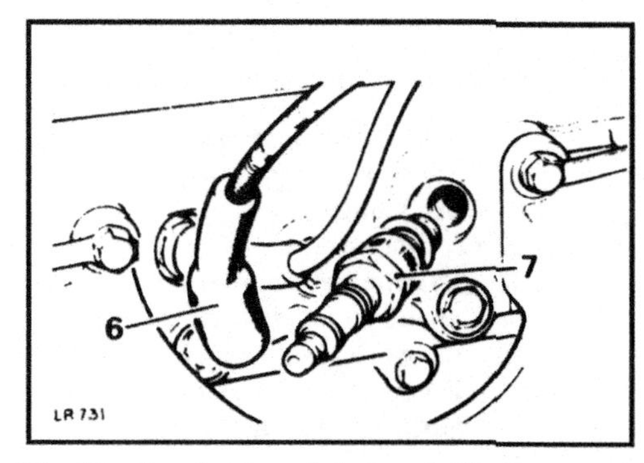

Fig. 29. Spark plug, right-hand side illustrated

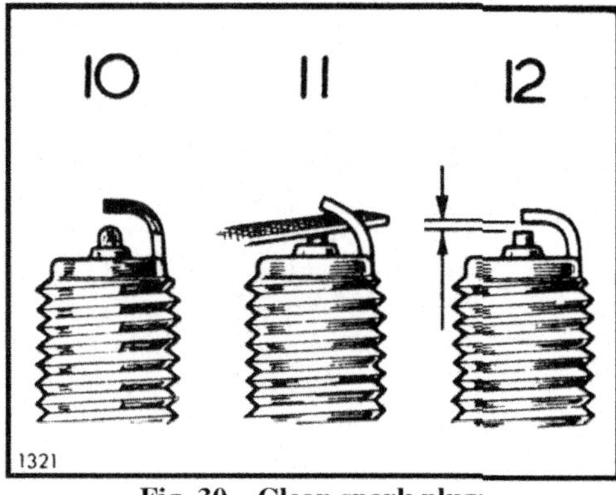

Fig. 30. Clean spark plugs

Distributor contact points—Check at free service 1.500 km (1,000 miles) and thereafter every 10.000 km (6,000 miles) or 6 months. Replace every 20.000 km (12,000 miles) or 12 months. Figs. 31 and 32.

To obtain satisfactory engine performance it is most important that the contact points are adjusted to the dwell angle which is 26° to 28°, using suitable workshop equipment. This work should be carried out by your local Rover Distributor or Dealer.

1. Remove distributor cap.
2. Remove the nut on the terminal block.
3. Lift off the spring and moving contact.
4. Remove adjustable contact, secured with a screw.
5. Add a smear of grease to contact pivot before fitting new contact points. Then carry out distributor maintenance followed by setting the ignition timing and dwell angle.

   However, when it becomes necessary to change the contact points and specialised checking equipment is not available, they may be adjusted either by the feeler gauge or alternatively the timing lamp method.

   Proceed as follows:

   Checking contact points

6. Turn the engine in direction of rotation until the contacts are fully open.
7. The clearance should be 0,35 to 0,40 mm (0.014 to 0.016 in.) with the feeler gauge a sliding fit between the contacts.
8. Adjust by turning the adjusting nut clockwise to increase gap or anti-clockwise to reduce gap.
9. Replace the distributor cap.

   At the first available opportunity after the contact points have been adjusted as detailed above they must be finally set to the dwell angle using specialised equipment.

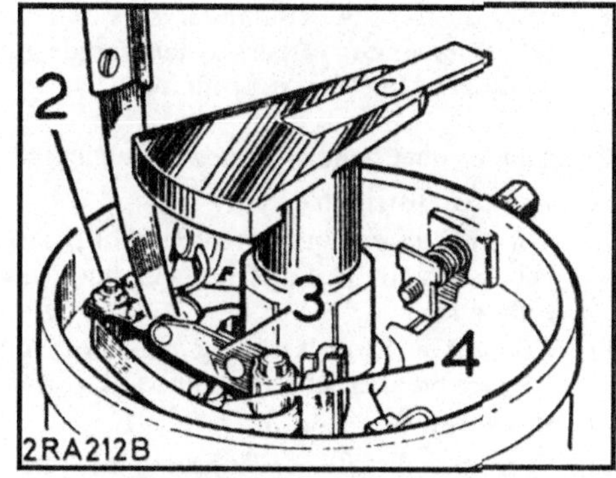

Fig. 31 Distributor contact points adjustment using the feeler gauge method

10. At the same time check the ignition timing which should be dynamically set to 5° ATDC at 750 revs/min maximum. When new contact points have been fitted, the dwell angle must be checked after a further 1.500 km (1,000 miles) running.

Checking contact points—*timing lamp method*

11. Remove distributor cap.
12. Turn the engine in the direction of rotation until the contact breaker heel is on the peak of number one cylinder cam. Points should be fully open.
13. Connect a 12 volt timing lamp, or suitable voltmeter, across the contact breaker lead terminal and a suitable earth point.
14. Switch on the ignition.
15. Turn the distributor adjusting nut **anti-clockwise** until the timing lamp goes out or there is no reading on the voltmeter.
16. Continue a further two turns of the adjuster in an anti-clockwise direction.
    During this operation the adjusting nut should be pressed inwards with the thumb to assist the helical return spring.
17. Slowly turn the adjusting nut **clockwise** until the timing lamp just comes on, or there is a voltage shown on the voltmeter.
18. Noting the position of the flats on the adjusting nut, continue in a clockwise rotation for a further **five** flats.
19. Remove timing lamp or voltmeter and switch off ignition.
20. Replace the distributor cap.
    At the first available opportunity after the contacts point have been adjusted as detailed above, they must be finally set to the dwell angle using specialised equipment.
21. At the same time, check the ignition timing, which should be dynamically set to 5° ATDC at 750 revs/min. maximum. When new contact points have been fitted, the dwell angle must be checked after a further 1.500 km (1,000 miles) running.

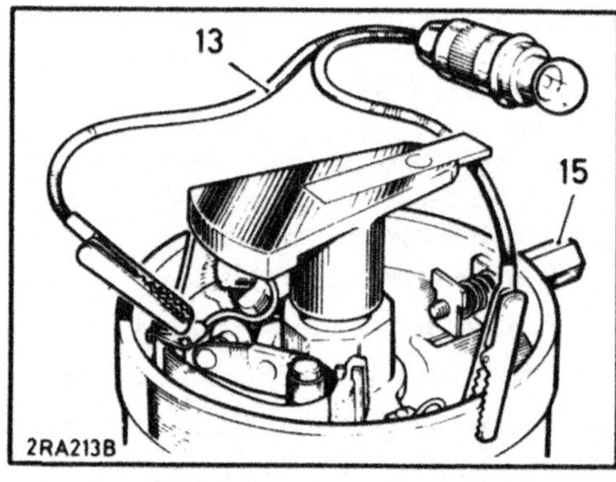

Fig. 32. Distributor contact points adjustment using the timing lamp method

Distributor maintenance—At free service 1.500 km (1,000 miles) and thereafter every 10.000 km (6,000 miles) or 6 months. Fig. 33.

Lubricate as follows:

1. Remove distributor cap.
2. Remove rotor arm.
3. Lightly smear the cam with clean engine oil.
4. Add a few drops of thin machine oil to lubricate the cam bearing and distributor shaft.
5. Wipe the inside and outside of the distributor cap with a soft dry cloth.
6. Ensure that the carbon brush works freely in its holder.
7. Replace rotor arm and distributor cap.

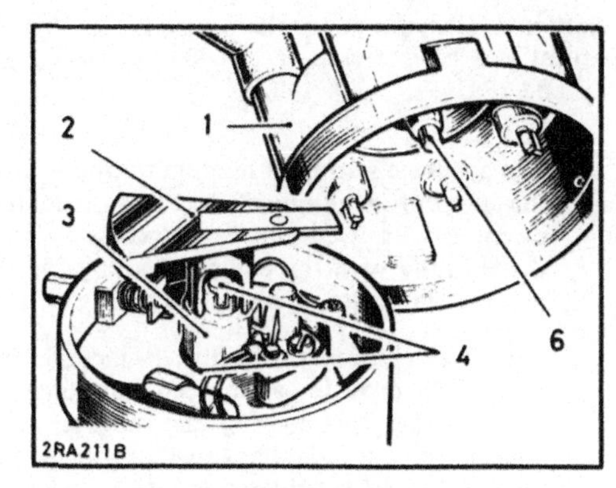

Fig. 33. Distributor maintenance

High tension leads—Check every 20.000 km (12,000 miles) or 12 months. Figs. 34 and 35.

1. A careful examination should be carried out on all high-tension leads, including the coil to distributor lead.
2. Look for any signs of insulation cracking or deterioration and corrosion at the end contacts.
3. Replace any faulty leads.
4. The correct sequence of plug leads is shown in Fig. 34.
   The numbers and letters in the circles indicate spark plug numbers and also the right-hand (RH) or left-hand (LH) bank of the engine to which the leads go.
5. High tension leads must be replaced in the correct relationship to each other, as well as ensuring correct firing order. Failure to do this will result in cross firing.
   The numbers in the arrowed circles, Fig. 35, show the plug lead numbers.
6. Check for loose or broken HT leads, cleats and clips.
7. Check locating clips fixed to rocker cover.

Note: The electrical leads to the ignition coil are fitted with male and female connectors; ensure that they are fitted to the correct blade on the coil.

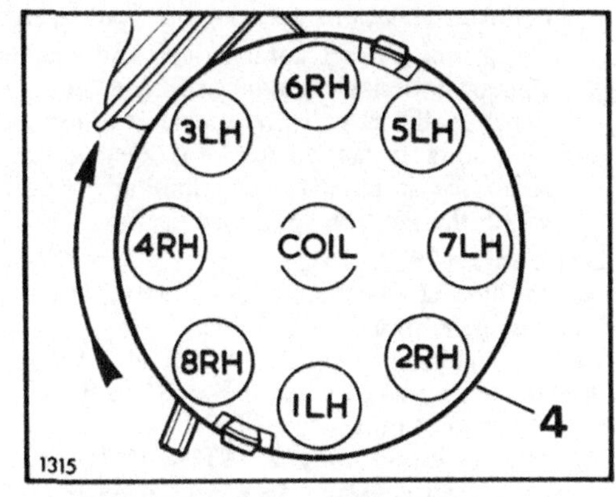

Fig. 34. Sequence of distributor leads

Setting dwell angle and ignition timing—At free service 1.500 km (1,000 miles) and thereafter every 10.000 km (6,000 miles) or 6 months. Figs. 36 and 37.

1. The accurate setting of ignition timing is of extreme importance, and the correct functioning of the emission control system relies to a large extent on its accuracy. It is necessary to set the ignition timing dynamically with the engine at idling speed. It is obvious therefore that this work should be carried out by a Rover Distributor or Dealer.

2. A special ignition distributor is included in the specification. The distributor provides a retarded ignition setting at the lower speed range whilst maintaining the normal advance characteristics at higher engine speeds. The distributor, together with the other modifications embodied, reduces exhaust emissions to an acceptable level.

   Failure to set ignition timing correctly, as subsequently described, will almost certainly result in the vehicle failing to comply with emission control regulations and can also lead to engine damage.

   To ensure correct combustion, and therefore compliance with the exhaust emission regulations, it is essential that the ignition timing is dynamically set with the engine at idling speed. This requires the use of a suitable tachometer, for determining the engine speed, and a stroboscopic lamp for determining the points in the engine cycle at which the ignition sparks occur.

Dwell angle: 26° to 28°.
Ignition timing when using fuel of 91 to 93 octane rating: †
Static ignition timing: 7° BTDC.
Dynamic ignition timing: 5° ATDC at 750 revs/min./maximum.
Ignition timing when using fuel of 85 to 91 octane rating: †
Static ignition timing: 4° BTDC.
Dynamic ignition timing: 8° ATDC at 750 revs/min./maximum.

†See General Data, Section 9, engine details for alternative ignition timing specified for non-emission engines

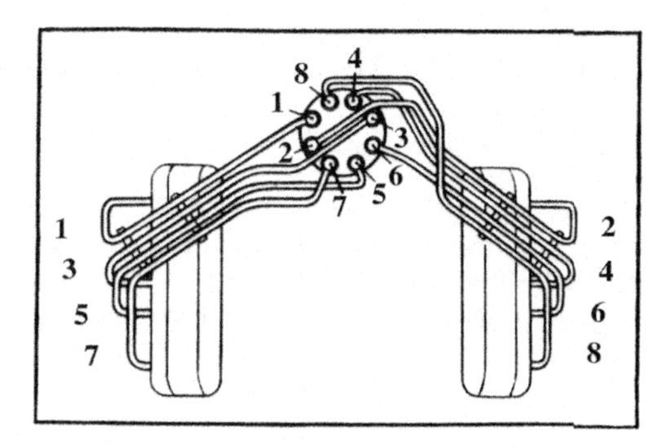

Fig. 35. Layout of high-tension leads

Carry out item 3 only if distributor has been disturbed.

3. Set ignition timing statically to 7° BTDC or 4° BTDC as applicable, prior to the engine being run, by the basic timing lamp method. (This sequence is to give only an approximation in order that the engine may be run. The engine must not be started after distributor replacement until this check has been carried out).

4. Set dwell angle as follows:

5. Start engine and set to an idling speed of 750 revs/min. maximum.

6. Set selector knob to 'calibrate' position on the tach/dwell meter. Adjust calibration knob to give a zero reading on the meter.

7. Couple meter to engine following manufacturer's instructions.

8. Set selector knob to 8 cylinder position and tach/dwell selector knob to 'dwell'. Adjust distributor dwell angle by turning the hexagon-headed adjustment screw on the distributor until the meter reads 26° to 28°.

9. Uncouple tach/dwell meter.
Care should be taken to switch the tach/dwell meter selector switch to the 'off' position after use, otherwise battery life will be impaired.

Set ignition timing as follows:

10. Couple a stroboscopic timing lamp to the engine following the manufacturer's instructions, with the high tension lead attached into No. 1 cylinder plug lead.

    Note: The two vacuum pipes must not be disconnected from the distributor vacuum capsule.

11. Start the engine and set to an idling speed of 750 revs/min maximum. using an accurate tachometer.

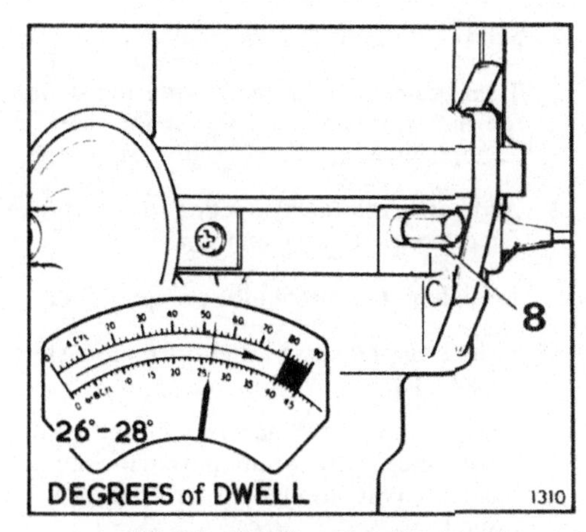

Fig. 36. Dwell angle adjustment

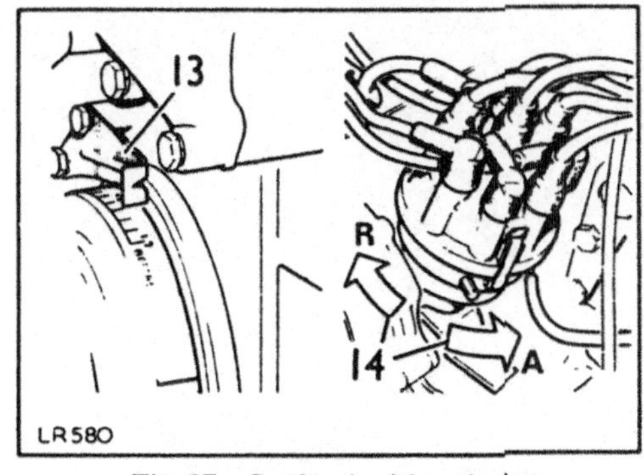

Fig. 37. Setting ignition timing

12. Slacken distributor clamping bolt.

13. Turn the distributor body until the stroboscopic lamp synchronises the timing pointer and the timing mark at 5° ATDC or 8° ATDC, as applicable, on the vibration damper rim.

14. Arrow (R) indicates direction to retard ignition. Arrow (A) indicates. direction to advance ignition.

15. Re-tighten the distributor clamping bolt.

16. Switch off the engine and disconnect stroboscopic timing lamp and tachometer.

    Note. Engine speed accuracy during ignition timing is of paramount importance. Any variation from the required 750 revs/min maximum, particularly in an upward direction, will lead to wrongly set ignition timing.

Range Rover maintenance location points for:

38 Fan belt adjustment
39 Power steering pump belt adjustment (as applicable)
40 Screen washer reservoir
41 Fuel filter element
42 Engine flame traps
43 Engine oil filler cap and oil level disptick
44 Power steering fluid reservoir (as applicable)

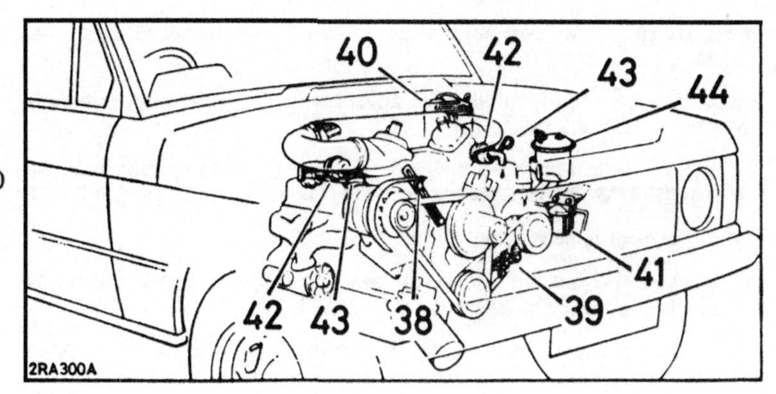

Fan belt adjustment—Check every maintenance inspection. Fig. 38.

Check adjustment again whenever a new fan belt is fitted, after approximately 1.500 km (1,000 miles) running.

1. Check by thumb pressure between alternator and crankshaft pulleys. Movement should be: 11 to 14 mm (0.437 to 0.562 in.).

   If necessary adjust as follows:

2. Slacken the bolts securing the alternator to the mounting bracket.
3. Slacken the fixings at the top and bottom of the adjustment link.
4. Pivot the alternator inwards or outwards as necessary and adjust until the correct tension is obtained, tighten the bolt at the top of the adjustment link.
5. Finally tighten the nut securing the bottom of the adjustment link and the two mounting bracket bolts.

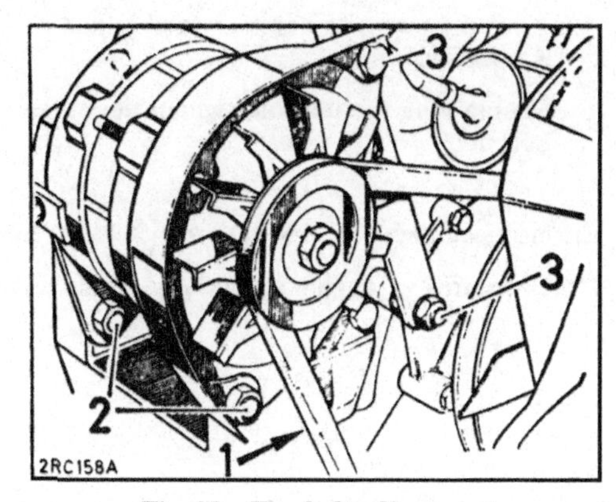

Fig. 28. Fan belt adjustment

Power steering pump belt adjustment, —Every maintenance inspection. Fig. 39.

Whenever a new belt is fitted check adjustment again after approximately 1.500 km (1,000 miles) running.

Check by thumb pressure the belt tension between the crankshaft and pump pulley. Movement should be 11 to 14 mm (0.437 to 0.562 in.).

If adjustment is necessary:

1. Slacken the nut on the pivot bolt securing the pump mounting bracket to the cylinder head.
2. Slacken the bolt securing the pump lower bracket to the slotted adjustment link.
3. Slacken the bolt securing the slotted adjustment link to the support bracket mounted on the water pump cover.
4. Pivot the pump as necessary and adjust until the correct belt tension is obtained.
5. Maintaining the tension, tighten the pump adjusting bolts and pivot bolt nut.

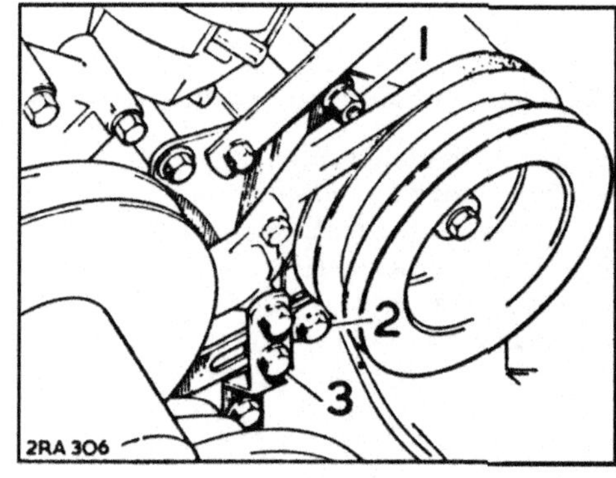

Fig. 39. Power steering pump belt adjustment

Engine mountings—At free service 1.500 km (1,000 miles) only.

Check security of engine mountings; rectify as necessary.

Water level, windscreen and rear screen washer reservoir—Every 1.000 km (750 miles) and at every maintenance inspection. Fig. 40.

The windscreen/rear screen reservoir is located on the left-hand bulkhead.

1. Remove either reservoir cap by turning anti-clockwise.
2. Top up reservoir to within approximately 25 mm (1 in.) below bottom of filler neck.
3. Use a screen washer solvent in the reservoir, this will assist in removing mud, flies and road film.
4. In cold weather, to prevent freezing of the water, add 'Isopropyl Alcohol' or methylated spirits.

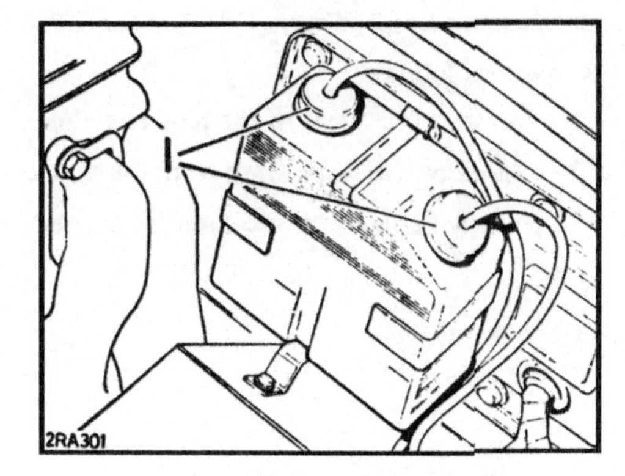

Fig. 40. Screen washer reservoir

Fuel filter element—Replace at free service 1.500 km (1,000 miles) and thereafter every 20.000 km (12,000 miles) or 12 months. Fig. 41.

The element provides a filter between the pump and carburetter and is located on the front LH wing.

Replace as follows:

1. Unscrew the centre bolt.
2. Withdraw the filter bowl.
3. Remove the small sealing ring and remove element.
4. Withdraw the large sealing ring from the underside of the filter body.
5. Discard the old element and replace with a new unit.
6. Ensure that the centre and top sealing rings are in good condition and replace as necessary.
7. Fit new element, small hole downwards.
8. Refit sealing rings.
9. Replace filter bowl and tighten the centre bolt.

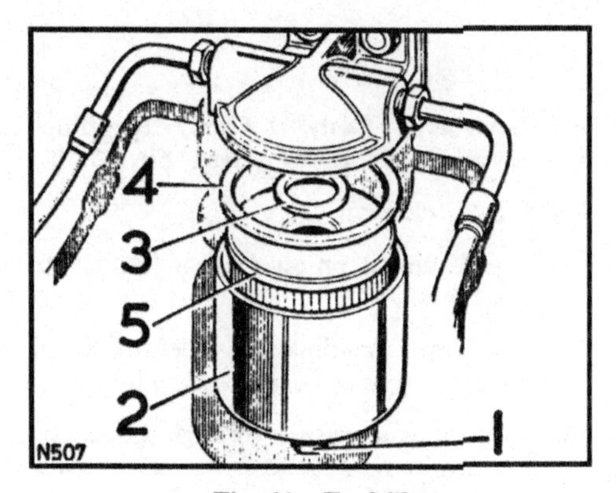

Fig. 41. Fuel filter

Engine flame traps—Every 20.000 km (12,000 miles) or 12 months. **Fig. 42.**

Replace as follows:

1. Pull the hoses clear of the retaining clips.
2. Withdraw the flame traps from the hoses.
3. Fit new flame traps into hoses and replace hoses to clips.

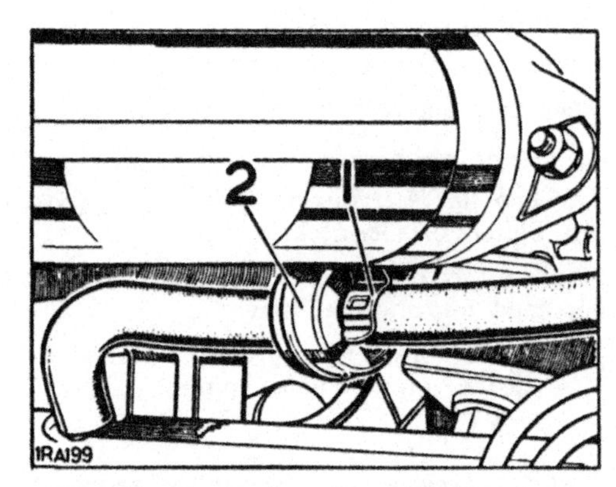

Fig. 42. Engine flame trap, RH illustrated

Engine oil level—Daily or weekly depending on operating conditions and at every maintenance inspection. Fig. 43.

Proceed as follows:

1. Stand the car on level ground and allow the oil to drain back into the sump.
2. Withdraw the dipstick at left-hand side of engine; wipe it clean, re-insert to its full depth and remove a second time to take the reading.
3. Add oil as necessary through the screw-on filler cap marked 'engine oil' on the right-hand front rocker cover. Never fill above the 'High' mark.

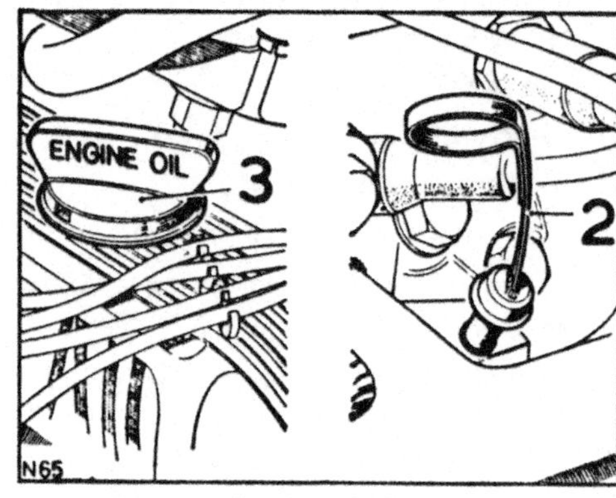

Fig. 43. Engine oil filler cap and level dipstick

Power steering fluid reservoir—At 1.000 km (750 miles) and at every maintenance inspection. Fig. 44.

The power steering units are lubricated by the operating fluid. The only lubrication required is to check the reservoir level as follows:

1. Unscrew the fluid reservoir cap.
2. Check that the fluid is up to the mark on the dipstick.
3. If necessary, top up using one of the recommended grades of fluid.

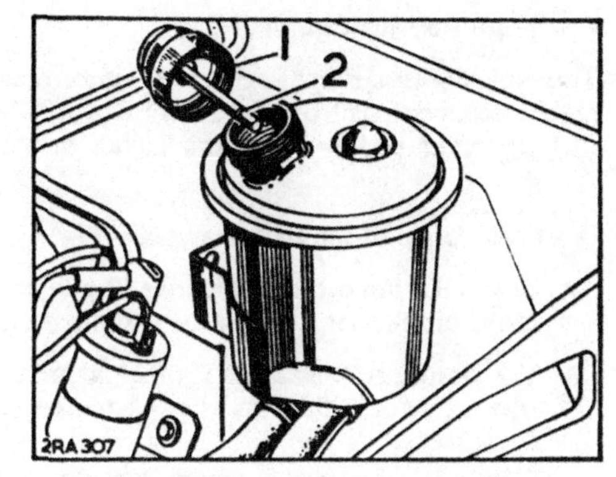

Fig. 44. Power steering fluid reservoir

Power steering—Every 5.000 km (3,000 miles) or 3 months.

Check power steering for oil leaks at hose connections, oil seals and joint faces on steering unit and power steering pump. Rectify as necessary.

Range Rover maintenance location points for:

45 Crankcase emission control system
46 Air intake temperature control system
47 Air cleaner and silencer flap valve
48 Air cleaner intake mixing flap valve
49 Air cleaner removal
50 Air cleaner element replacement
51 Breather filter for engine

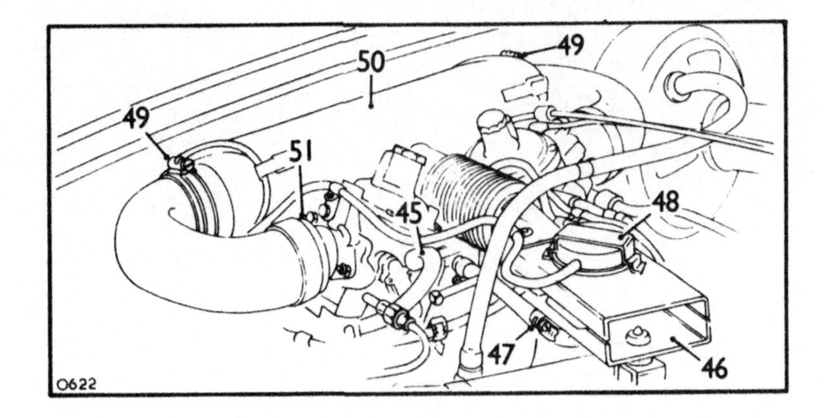

Crankcase emission control

To comply with current regulations concerning engine emission control, crankcase emissions from the Range Rover V8 engine are vented into the carburetters to be burnt with the fuel/air mixture.

Brief description of the control system—Fig. 45.

1. The breathing cycle is performed by tapping clean air from the rear of the air cleaner, then to the crankcase via a hose and filter.
2. The crankcase fumes rise via the pushrod tubes to the rocker covers where they are then transferred to the carburetters via hoses and flame traps.

   Finally the fumes are drawn into the engine to be burnt with the fuel/air mixture.

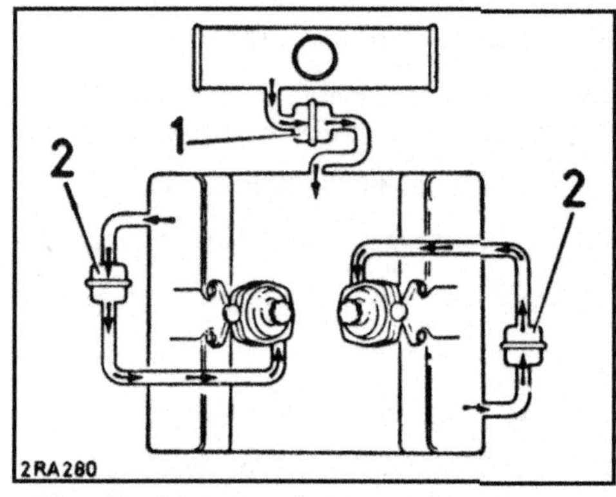

Fig. 45. Diagramatic layout of crankcase emission control system

Air intake temperature control—Figs. 46 and 47.

To enable the engine to operate on the most efficient air/fuel ratio, a system is incorporated which is designed to achieve an air intake temperature of 38°C (100°F) as soon as possible and maintain this temperature whilst ambient conditions are below 38°C (100°F).

The system comprises:

1. A hot box surrounding the right-hand exhaust manifold.

2. A vacuum operated thermostatically controlled flap valve in the air cleaner and silencer intake.

3. The flap valve controls the source of the intake air supply which may be warm air drawn entirely from the hot box or cold air drawn from the under bonnet area or a combination of both.

4. The hot box is connected via a hose to the flap valve in the air intake.

5. The temperature sensing device is situated in the air cleaner on the clean side of the right-hand element.

6. A pipe from the manifold is attached to the temperature sensing device via a non-return valve.

7. From the other side of the temperature sensing device is a pipe connecting the vacuum capsule operating the flap valve.

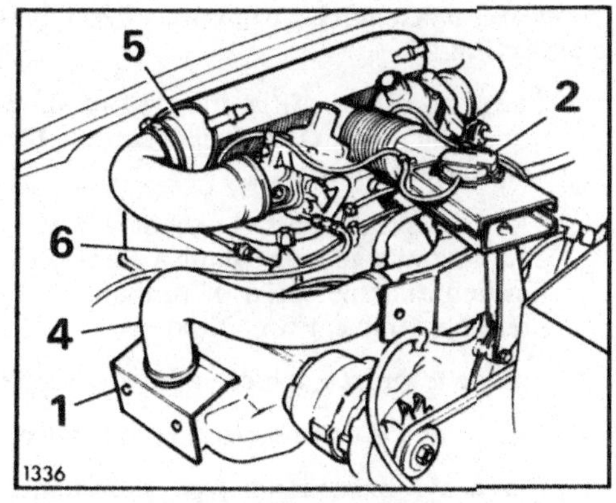

Fig. 46. Layout of air intake temperature control system

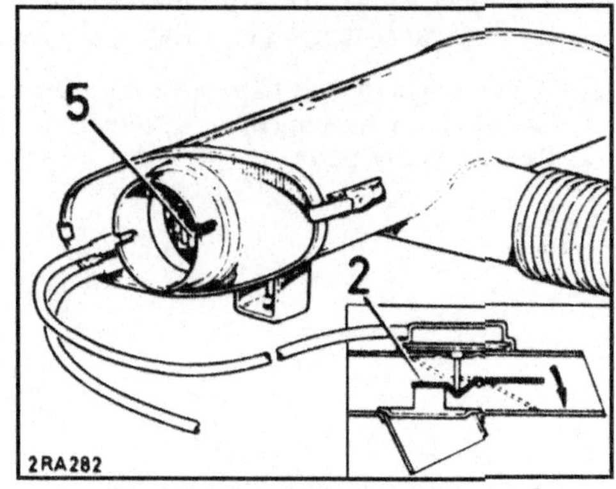

Fig. 47. Layout of air cleaner and silence flap value

Air cleaner intake mixing flap valve—Every 20.000 km (12,000 miles) or 12 months. Fig. 48.

1. Check operation of the mixing flap valve in air cleaner by starting up engine from cold and observing the flap valve as the engine temperature rises.
2. The valve should start to open slowly within a few minutes of starting and continue to open until a stabilised position is achieved. This position and the speed of operation will be entirely dependent on prevailing ambient conditions.
3. Failure to operate indicates failure of the flap valve vacuum capsule.
4. Failure of the thermostatically controlled vacuum switch or both.
5. Check by connecting a pipe, shown in dotted line, one end direct from manifold tapping.
6. The other end to the flap valve vacuum capsule, thereby by-passing the air cleaner temperature sensing device.
7. If movement of the flap valve is evident the temperature sensor is faulty. If no movement is detected, the vacuum capsule is faulty. Replace faulty parts.

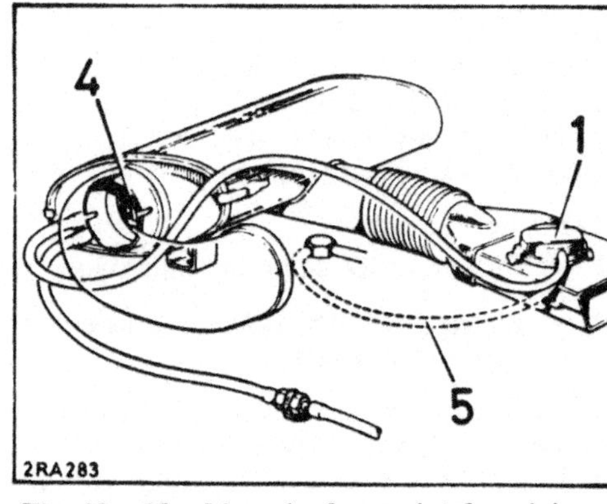

Fig. 48. Checking air cleaner intake mixing flap valve

Air cleaner element replacement—Every 20.000 km (12,000 miles) or 12 months. Figs. 49 and 50.

Attention to the air cleaner is extremely important. Replace elements every 10.000 km (6,000 miles) or 6 months under severe dusty conditions, as performance will be seriously affected if the engine is run with an excessive amount of dust or industrial deposits in the elements.

For air cleaner removal proceed as follows, following the instructions applicable to the type of air cleaner fitted:

1. Slacken the clip retaining the advance/retard vacuum pipes from the air intake and release pipes from intake.
2. Slacken the clip retaining hose air cleaner to temperature sensing device from air intake and remove pipe from flap valve on air intake.
3. Slacken the hose clip attaching warm air intake hose to air intake.
4. Withdraw the air intake from the steady post and hoses.
5. Slacken the clips securing the air cleaner elbows and withdraw elbows.
6. Remove the air cleaner from the retaining posts by lifting and easing forward.
7. At the same time disconnect the hose engine breather filter to air cleaner. Place air cleaner to one side.
8. Remove hose with the non-return valve from the manifold.

The air cleaner can now be completely removed.

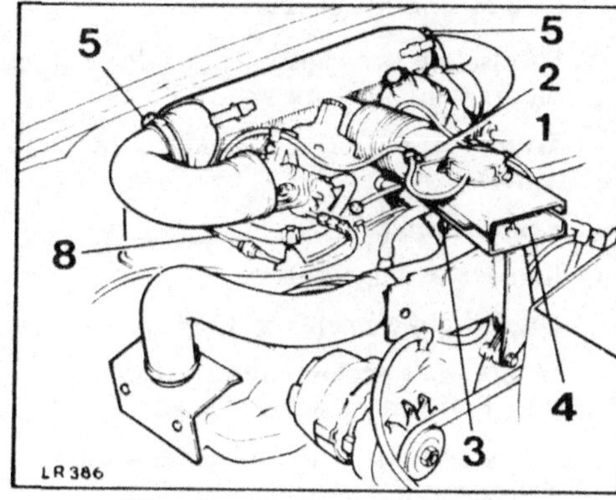

Fig. 49. Air cleaner removal

For air cleaner element replacement:

1. Release the two clips at each side of air cleaner casing and withdraw the frames and elements.
2. To replace the elements remove the screw and washer on the frame carrier.
3. Remove end cap.
4. Remove sealing washers.
5. Discard old elements and replace with new units.
6. Ensure that sealing washers on frame and end cap are in good condition and correctly located.
7. Check condition of rubber seals on end of carrier frame. Replace if necessary.
8. Reassemble elements to air cleaner and air cleaner to engine by reversing the removal procedure as follows:
9. Fit carrier frames in air cleaner casing and secure with the two clips at each side.
10. Reconnect hose with non-return valve to the manifold connection.
11. Place air cleaner on to the retaining posts.
12. At the same time reconnect the breather filter hose to the base of the air cleaner.
13. Refit the air cleaner elbows and tighten clips.
14. Replace air intake on to steady post and reconnect air cleaner and warm air intake hoses. Tighten clip.
15. Reconnect pipe, air cleaner to temperature sensor and vacuum advance/retard pipe to air intake. Position pipes in retaining clips on air intake and tighten clips.

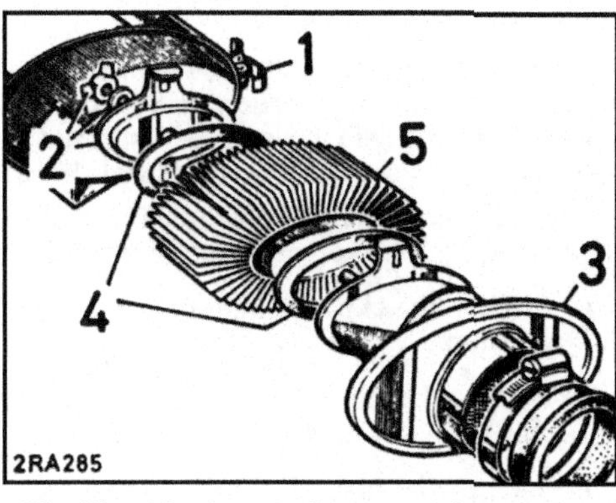

Fig. 50. Air cleaner element replacement

Engine breather filter—Every 40.000 km (24,000 miles) or 24 months.

Replace as follows:

1. Remove the air cleaner as detailed under 'Air cleaner'.
2. Withdraw rear hose from the filter.
3. Slacken the filter clip.
4. Withdraw the filter from the clip and front hose.
5. Fit new filter with end marked 'IN' facing forward. Alternatively, if the filter is marked with arrows, they must point rearwards. Refit hoses and tighten clip.

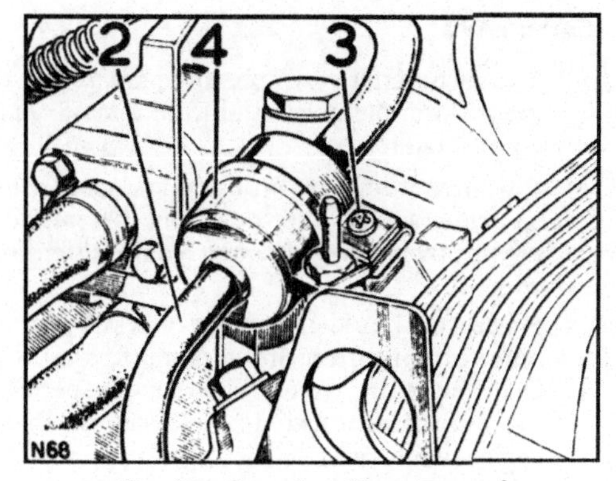

Fig. 51. Breather filter for engine

Carburetters

1. The carburetters are specially prepared instruments manufactured to extra close tolerances and form a major part of the exhaust emission control equipment.

   Carburetter mixture ratio and idle speed settings are pre-set and sealed at manufacture and must not be interfered with. Under normal circumstances they do not require attention except at major engine overhaul.

   However, should it become necessary to check any aspect of carburetter adjustment the work must be carried out by a qualified Rover Distributor or Dealer, who has the specialised equipment needed to carry out adjustments to the close limits necessary to ensure that the engine conforms to the legal requirements in respect of exhaust emission.

   Under no circumstances must the mixture be disturbed, as this would almost certainly result in the vehicle failing to meet with legal requirements in respect of air pollution.

   The idle adjuster screws have seals fitted to prevent unauthorised adjustment or movement due to vibration.

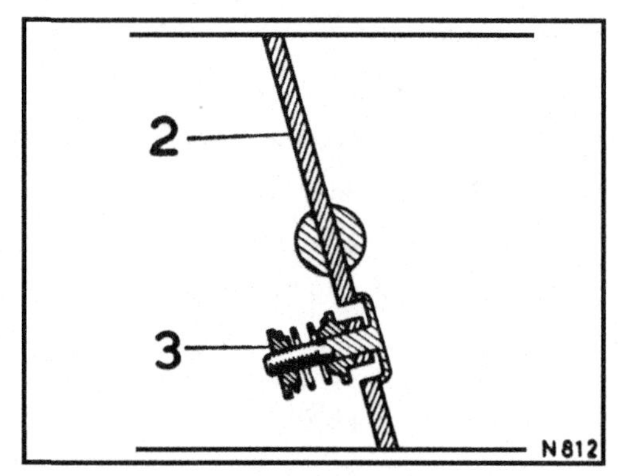

Fig. 52. Throttle butterfly, low manifold depression

Carburetter throttle butterfly. Figs. 52 and 53.

2. The throttle butterfly has a spring loaded poppet valve.
3. With low manifold depression, the valve remains closed.
4. At high manifold depression conditions, Fig. 53, that is over-run at closed throttle, the valve opens and prevents incorrect combustion of fuel by supplementing the volume of fuel/air mixture; this together with a vacuum retarded ignition setting maintains correct combustion.

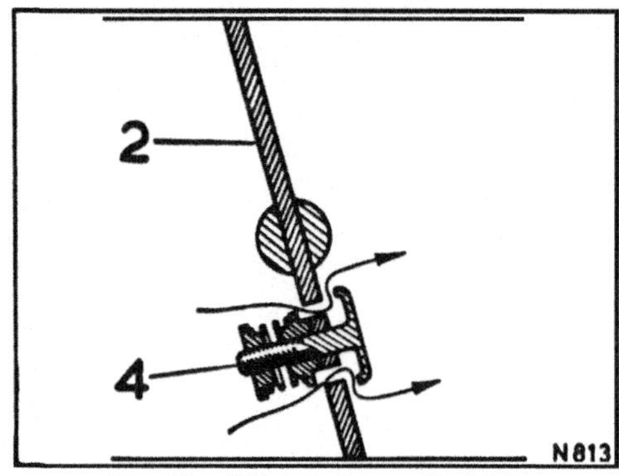

Fig. 53. Throttle butterfly, high manifold depression

Spring loaded carburetter needles—Fig. 54.

1. Each carburetter needle is spring loaded.
2. The needle is biased by the spring against the retainer.
3. This maintains the needle in its correct relationship with the carburetter jet thus improving the control of emission.

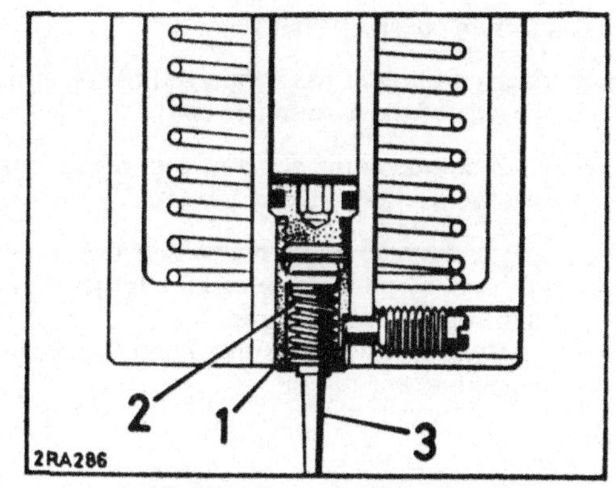

Fig. 54. Cross-section view of spring-loaded needle

Fuel deflector—Fig. 55.

1. A fuel deflector is fitted between the carburetter insulating block and inlet manifold.
2. The deflector takes the form of a sheet of metal pressed out to give a hole with inward facing teeth through which the fuel/air mixture passes and is atomised.
3. The purpose of the teeth is to prevent wet fuel accumulating on the manifold walls and thus allowing the engine to function satisfactorily on leaner mixtures.

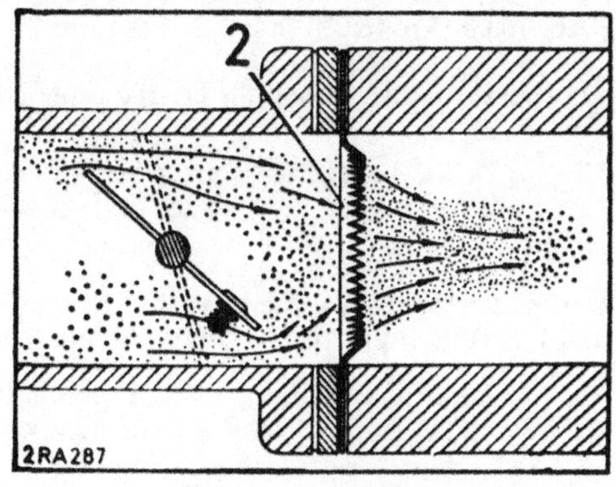

Fig. 55. Fuel deflector

Temperature compensator—Fig. 56.

1. Each carburetter has an integral temperature compensator situated at the side of the carburetter body.
2. Each compensator contains a bi-metallic blade which is sensitive to carburetter body temperature.
3. The bi-metallic blade regulates a pre-set tapered plug and allows an air bleed by-passing the carburetter jet thus giving weaker mixtures at high engine temperatures.
   The system gives sensitive mixture control over a wide range of temperatures.

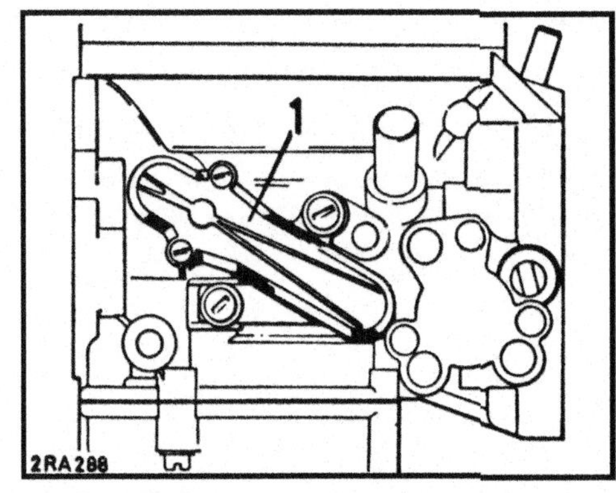

Fig. 56. Carburetter temperture compensator

Accelerator linkage—At free service 1.500 km (1,000 miles) and thereafter every 10.000 km (6,000 miles) or 6 months.

1. Lubricate the accelerator linkage using clean engine oil.
2. Check the linkage for correct operation and ensure that there is no tendency to stick. Badly worn parts should be replaced.

Carburetter hydraulic dampers—At free service 1.500 km (1,000 miles and thereafter every 10.000 km (6,000 miles) or 6 months. Fig. 57.

1. Unscrew the cap on top of each suction chamber; withdraw cap and hydraulic damper. Replenish the damper reservoir as necessary with SAE 20 oil to within about 12 mm ($\frac{1}{2}$ in.) from the top of the tube.
2. Then replace cap and hydraulic damper.

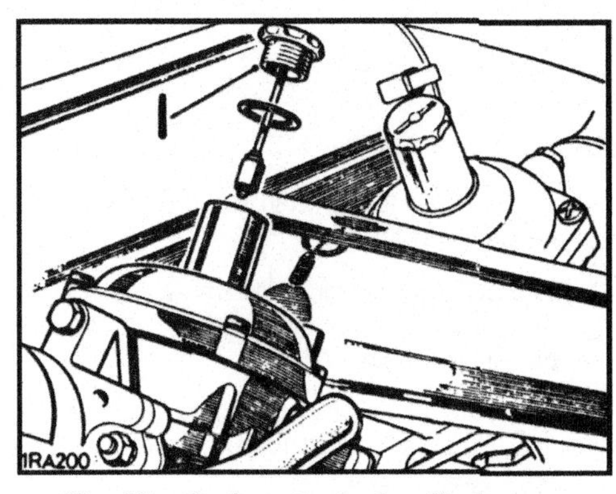

Fig. 57. Carburetter hydraulic damper

Carburetter choke adjuster—Adjust to suit climatic conditions.

1. For starting at temperatures down to —18 °C (0 °F) push and turn the spring-loaded choke adjustment screw so that the peg is at right-angles to the slot as illustrated. Leave in this position.
   When starting at temperatures below —18 °C (0 °F) turn the screw until peg is recessed in slot.

Fig. 57a. Choke adjustment screw

Road test—At free service 1.500 km (1,000 miles) and thereafter every 10.000 km (6,000 miles) or 6 months.

Give the vehicle a thorough road test and carry out any further adjustments required.

Check operation of all instruments and warning lights in facia panel.

After test, check for oil, fuel, fluid or grease leaks at all plugs, flanges, joints and unions.

Ensure that controls, door handles, steering wheel, etc. are clean and free from grease.

Preventive maintenance

In addition to the recommended periodical inspection of brake components it is advisable as the car ages, and as a precaution against the effects of wear and deterioration, to make a more searching inspection and renew parts as necessary.

It is recommended that:

1. Disc brake pads, drum brake linings, hoses and pipes should be examined at intervals no greater than those laid down in the Maintenance Schedules of the Handbook.
2. Brake fluid should be changed completely every 18 months or 18,000 miles (30.000 km) whichever is the sooner.
3. All fluid seals in the hydraulic system and all flexible hoses should be renewed every 3 years or 37,500 miles (60,000 km) whichever is the sooner. At the same time the working surfaces of the piston and the

bores of the master cylinders, wheel cylinders, and other slave cylinders should be examined and new parts fitted where necessary.

Care must be taken always to observe the following points:

(a) At all times use the recommended brake fluid.

(b) Never leave fluid in unsealed containers. It absorbs moisture quickly and can be dangerous if used in your braking system in this condition.

(c) Fluid drained from the system or used for bleeding is best discarded.

(d) The necessity for absolute cleanliness throughout cannot be overemphasized.

Replacing brake-shoes

When it becomes necessary to renew the brake-shoes, it is essential that only genuine shoes, with the correct grade of lining are used. Always fit new shoes as complete sets, never individually.

Replacement brake-shoes are obtainable from your Dealer.

Bleeding the brake system—Figs. 58 and 59.

If the brakes feel spongy, this may be caused by air in the hydraulic system. This air must be removed by bleeding the hydraulic system at the disc cylinders; one bleed point at each side on the rear, and three at each side on the front.

The following additional points should be noted when bleeding the dual system. Varying brake pedal travel will be experienced depending upon the degree of bleeding required. Bleeding the primary system, that is, both rear brakes and half the front brakes with the secondary system fully operational, almost full brake pedal travel can be used. When bleeding the secondary system, that is, half the front brakes only, with the primary system fully operational, approximately half the total brake pedal travel can be used.

Important: If bleeding the secondary system only, commence with the front caliper furthest from the master cylinder, and bleed from the screw on the same side as the fluid inlet pipes, then close the screw, and bleed from the screw on the opposite side on the same caliper. Repeat for the other front caliper.

An advantage with the dual system is the ability to change brake components without having to bleed both systems. The only requirement to bleed the complete system is following the removal and refitting of the servo unit, the pressure failure switch, or complete disconnection and reconnection of the pipes.

To bleed the complete system, proceed as follows:

1. Attach a length of rubber tubing to the bleed screw on the rear left-hand caliper and place the lower end of the tube in a glass jar containing brake fluid.
2. Slacken the bleed screw.
3. When the fluid appears in the jar, commence pumping the brake pedal slowly; pause at least five seconds at each end of the return stroke to allow the master cylinder piston to recuperate. Continue pumping until the fluid issuing from the tube shows no signs of air bubbles when the tube is held below the surface of the fluid in the jar.
4. Hold the tube under the fluid surface, and with the foot brake fully depressed, tighten the bleed screw and replace the dust cap.
5. Repeat this procedure for the rear right-hand caliper.
6. Attach a bleed tube to the primary bleed screw on the front caliper **furthest away** from the master cylinder.
7. Attach a second bleed tube to the secondary bleed screw on the same side of the caliper as the primary bleed screw, using two separate bleed jars.
8. Slacken both bleed screws.
9. When the fluid appears in the bleed jars, commence pumping the brake pedal slowly, pausing at each end of the return stroke, to allow the piston to recuperate, until fluid being expelled is free of air in both jars.

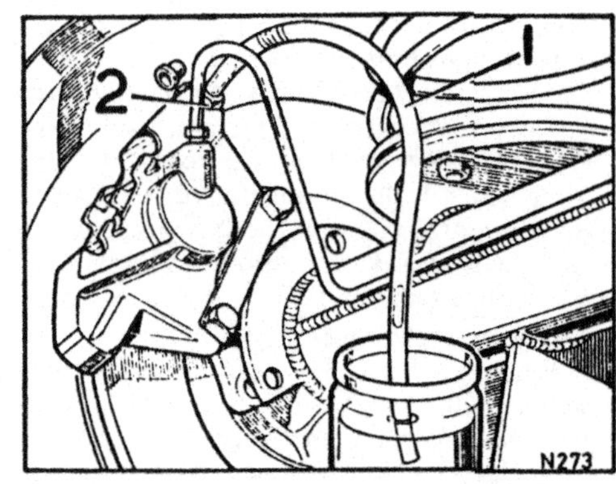

Fig. 58. Rear brake bleed screw

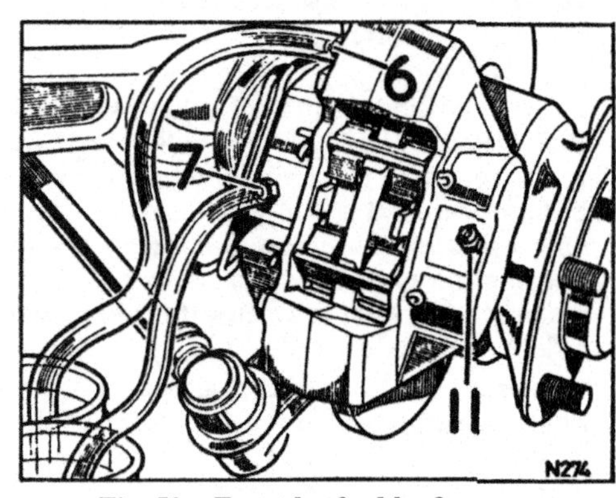

Fig. 59. Front brake bleed screws

10. Hold the tubes under the fluid surface and with the brake pedal fully depressed, tighten both bleed screws and replace both dust caps.
11. Attach a bleed tube to the remaining secondary bleed screw on the same caliper.
12. Slacken the bleed screw.
13. When the fluid appears in the jar commence pumping the brake pedal slowly, pausing at each end of the return stroke, to allow the piston to recuperate, until all air is expelled. Hold the tube under the fluid surface and with the brake pedal fully depressed, tighten the bleed screw and replace dust cap.
14. Repeat this procedure for the front caliper nearest to the master cylinder.
15. The fluid in the reservoir should be replenished throughout the operation to prevent another air lock being formed, using only new fluid of the recommended type, from sealed tins.
16. It will be obvious that the above operation requires two people.

Bulb Changing and Wheel Changing

7

This section of the book gives details of bulb changing, wheel changing and jacking, etc.

Headlamps—Fig. 60.

To replace headlamp bulb:

1. Prop open the bonnet. Two large clearance holes are provided, one on each side of the front valance, to give access to the respective bulb holders in the headlamp reflectors.

   Note: To obtain access to the right-hand clearance hole it will be necessary to remove the battery from the vehicle.
2. Disconnect the multi-plug lead.
3. Remove the rubber dust cover.
4. Release the bulb retaining spring clip.
5. Remove the faulty bulb and fit the correct 'Halogen' type. The bulb holder is keyed to facilitate fitting.

   IMPORTANT: Do not touch the quartz envelope of the bulb with the fingers. If contact is accidentally made wipe gently with methylated spirits.
6. Refit the bulb retaining spring clip rubber dust cover and multi-plug lead.
7. In the case of right-hand bulb replacement, re-fit the battery.

Side lamps—Fig. 61.

To replace side and flasher lamp bulbs:

1. Remove four posi-drive screws securing lens.
2. Withdraw lens and sealing rubber.
3. Replace bulb: top direction indicator flasher lamp bulb, bottom-side lamp bulb.
4. Refit lens and sealing washer; do not overtighten screws.

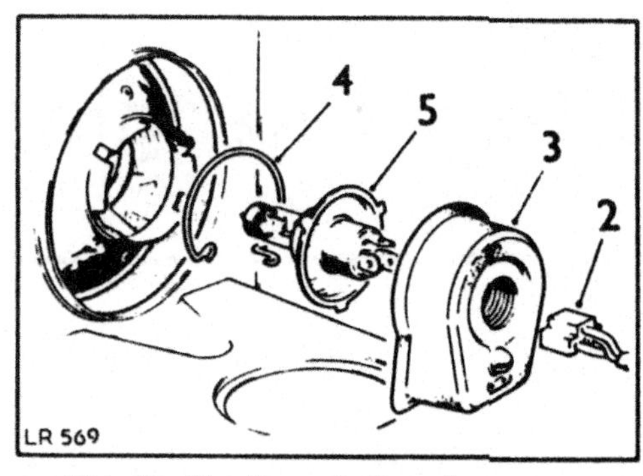

Fig. 60. Headlamp bulb replacement

Fig. 61. Side flasher lamp bulb replacement

Rear lamps—Fig. 62.

To replace flasher, tail, reverse and fog guard lamp bulbs.

1. Remove four screws retaining lens.
2. Remove lens.
3. Replace bulb: top, tail/stop lamp bulb.
4. Direction indicator flasher lamp bulb.
5. Centre, inner, reverse lamp bulb.
6. Bottom, fog guard lamp bulb.
7. Refit lens; do not overtighten screws.

Fig. 62. Rear lamp bulb replacement

Number plate illumination lamp—Fig. 63.

To replace number plate illumination bulb:

1. Remove two screws retaining lens hood.
2. Remove lens and sealing rubber.
3. Replace bulb as required.
4. Refit lens and sealing rubber.

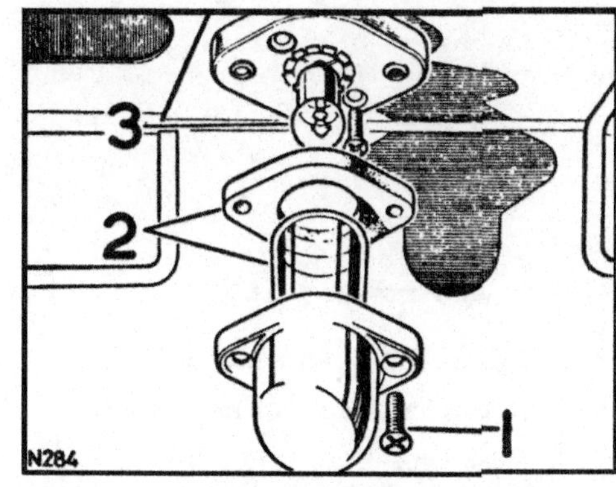

Fig. 63. Number plate illumination lamp

Warning light bulbs:

Rear fog guard lamps, differential lock, side lamps (and 'park' lamp where specified)—Fig. 64.

To replace a warning light bulb:

1. **Ease cover complete with bulb and bulb holder from the facia.**
2. **Withdraw the bulb holder from the cover and remove bayonet type bulb.**
3. **Renew bulb, replace bulb holder into the cover and replace cover. The cover is located in position by two small projections.**

Fig. 64. Differential lock switch illumination

Interior roof lamps—Fig. 65.

Two circular lamps, fore and aft in roof:

1. **Turn lens anti-clockwise and withdraw.**
2. **Replace bulb**
3. **Refit lens.**

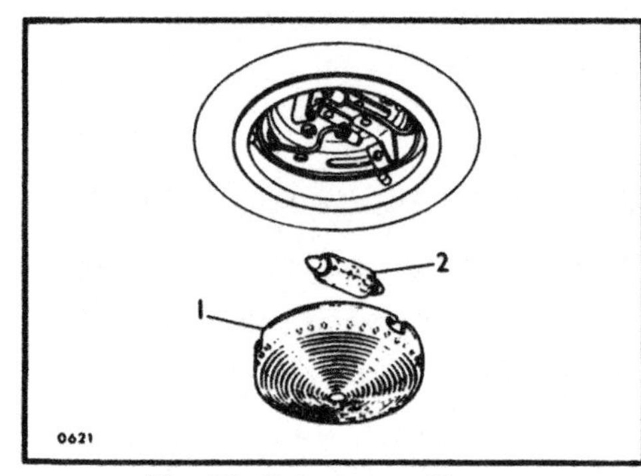

Fig. 65. Interior roof lamp

Warning and panel lights—Figs. 66 and 67.

To replace any of the warning light bulbs in the instrument binnacle:

1. **Release the instrument binnacle by pressing inwards with fingers at the bottom rear edge.**
2. **Lift binnacle over retaining clips.**
3. **Ease binnacle to one side for access to bulb holders.**
4. **Change bulb as required by pulling the bulb holder from its socket.**
5. **Replace bulb and reverse removal procedure.**

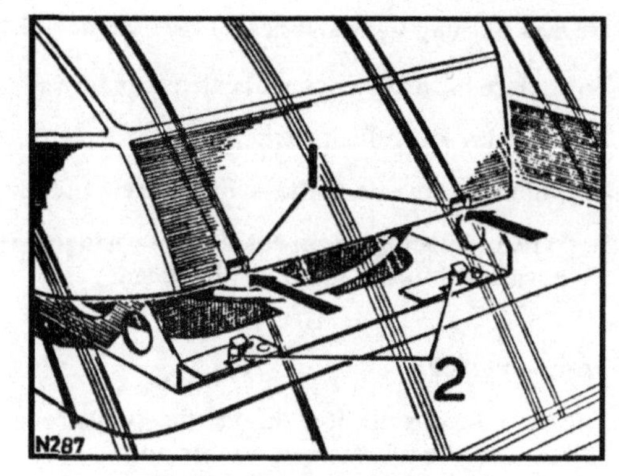

Fig. 66. Instrument binnacle

Direction indicator side repeater lamps, on front wings – Fig. 67a

1. Remove screw retaining the front of the lens.
2. Slide the lens forward to disengage it from the lamp base projection.
3. Remove the lens (and gasket if loose).
4. Renew the bulb.
5. Reset the gasket and replace the lens and fix by screw.

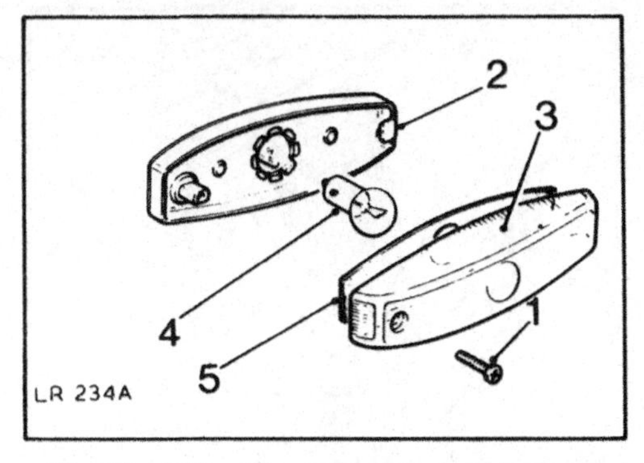

Fig. 67a. Direction indicator repeater lamp

Fig. 67. Warning lights, bulb replacement

Hazard warning light and heated rear screen switch—Fig. 68.

To replace hazard switch or heated rear screen switch warning light bulb:

1. Unscrew knob from switch.
2. Bulb can now be withdrawn and new bulb fitted.
3. When replacing knob, ensure that spring retaining the bulb is correctly positioned in the knob as illustrated.

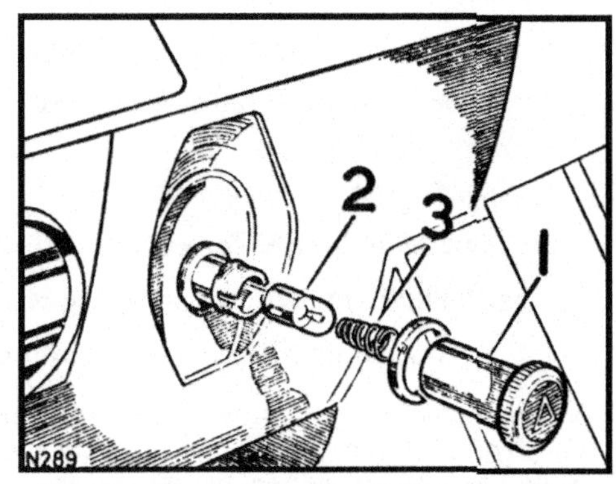

Fig. 68. Hazard warning light switch bulb replacement illustrated

Fuses—Fig. 69.

The fuse box is located under the bonnet on the left-hand bulkhead, adjacent to the windscreen washer bottle.

To replace fuse:

1. Pull off fuse box cover.
2. Replace fuse as required.
3. Two spare 35-amp fuses are clipped into the fuse box.

IMPORTANT

To ensure that regulations existing in certain countries concerning the hazard warning system are not contravened, it is most important that in the event of the 'Battery-Auxiliaries' fuse blowing for any reason, it is not removed and discarded until a replacement fuse can be obtained and fitted.

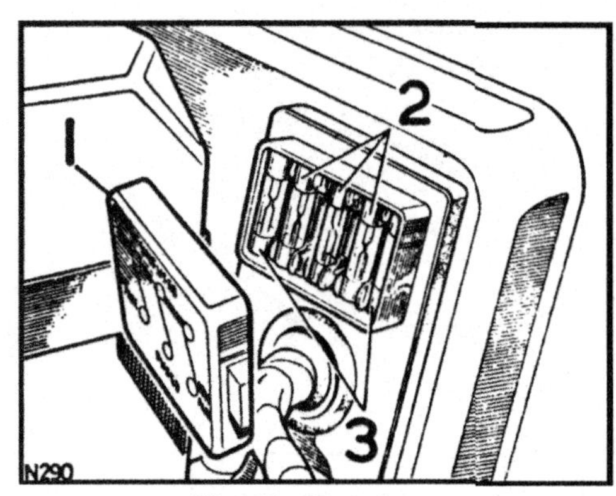

Fig. 69. Fuse box

Circuit diagram

Circuit diagram will be found at the end of the General Data Section 9.

Clock and bulb removal—Figs. 70 and 71.

1. Disconnect the battery.
2. Remove the grub screws retaining the four heater control knobs to levers, and remove knobs.
3. Remove the two screws securing the heater escutcheon plate, and remove plate.
4. Remove the four upper retaining screws securing the heater console to the facia panel.
5. Remove the centre face level louvre.
6. Lower the passenger glove box lid.
7. Remove the lower screw securing the heater console.
8. Ease the console forward to gain access to the clock.
9. Remove the feed wire at the '+' Lucar connector.
10. Remove the earth wire from the '—' Lucar connector.
11. Remove the clock illumination lead from the Lucar connector.
12. Remove the knurled nut and bracket securing the clock to the console and remove clock.
13. Replace bulb if required by pulling bulb and holder from clock.
14. Replace clock and heater console by reversal of the removal procedure. Set clock hands by pressing and rotating knob on clock face.
15. Reconnect battery.

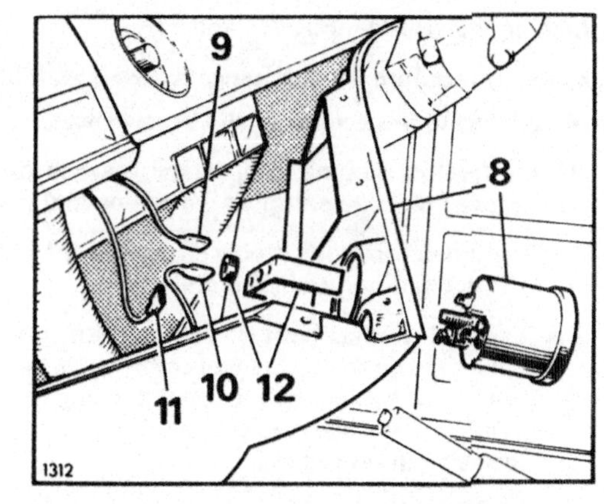

Fig. 70. Clock removal from fascia

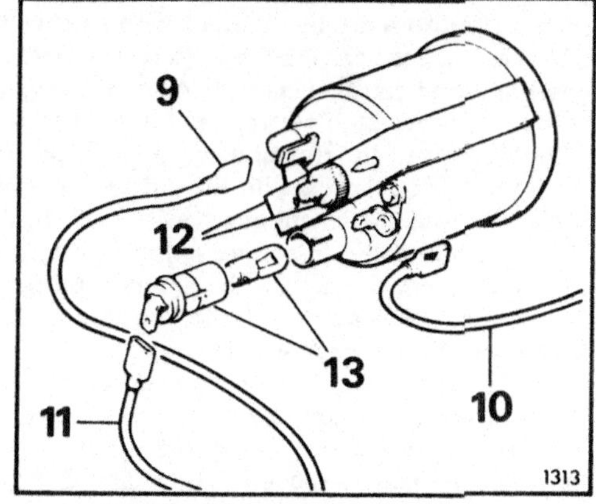

Fig. 71. Clock bulb removal

Jacking the vehicle—Figs. 72 and 73.

Jacking procedure for the Range Rover is as follows:

Wheels should be chocked in all circumstances.

1. On level or sloping ground, the gearbox differential lock should be engaged prior to stopping the engine and parking the vehicle.
2. The differential lock is only engaged if the warning light is illuminated with the ignition switched on.
3. If the vehicle has been stationary prior to the jacking operation, the differential lock may not operate when the switch is raised. In this case it will be necessary to start the engine, to create a vacuum and, perhaps drive the vehicle, until the warning light is illuminated. Then switch off engine.
4. Apply the handbrake.

Explanatory note: Owing to the fact that the vehicle is fitted with a transmission handbrake, this will not be operative if the differential lock has not been engaged and one or both rear wheels are jacked up, whilst either gearbox is in neutral. Therefore, to obtain engine braking, both gearboxes should be engaged in 1st gear and 'low' transfer respectively.

The design of the transmission is such that jacking up the rear wheels, whilst on a slope, even with the differential lock engaged, could result in limited vehicle movement as a result of the 'back-lash' in the transmission.

The handbrake is operative within transmission back-lash limits, if the rear wheels are to stay on the ground and one or both front wheels are jacked up, irrespective of the gearbox differential lock engagement. **Therefore always chock wheels.**

To jack up a front wheel: Jack up the corner of the vehicle by positioning the jack so that when raised, it will engage with the front axle casing immediately below the coil spring where it will be located between the flange at the end of the axle casing and a large bracket to which front suspension members are mounted, see Fig. 72.

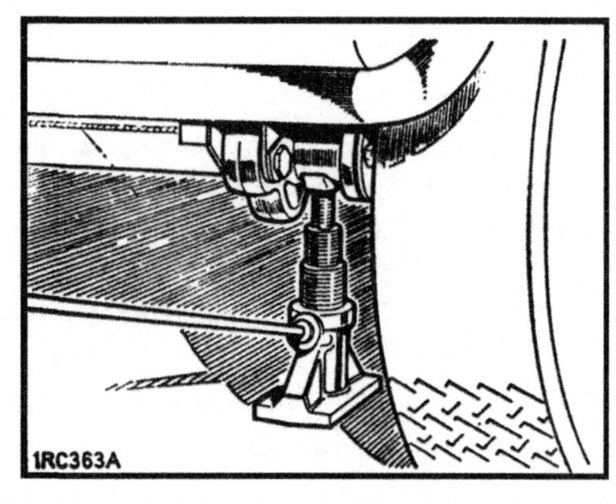

Fig. 72. Correct position for jack, front

To jack up a rear wheel: Jack up the corner of the vehicle by positioning the jack so that when raised it will engage with the rear axle casing immediately below the coil spring and as close to the shock absorber mounting bracket as possible, see Fig. 73.

Warning: It is unsafe to work under the vehicle using only the jack to support it. Always use stands or other suitable supports to provide adequate safety. Neglect of the jack may lead to difficulty in a roadside emergency. Examine the jack occasionally; clean and grease the thread to prevent the formation of rust.

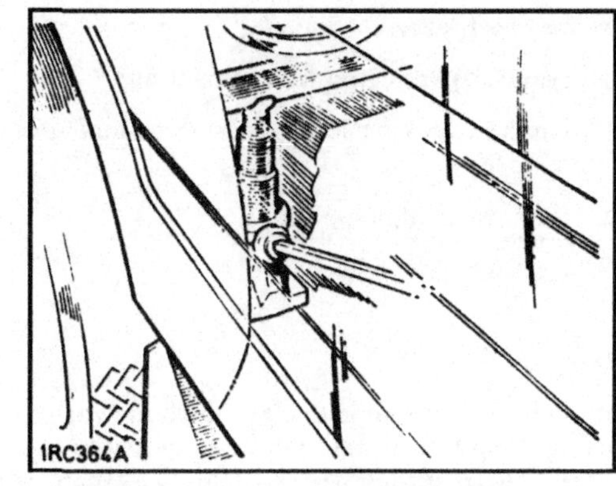

Fig. 73. Correct position for jack, rear

Wheel changing—Fig. 74.

1. Slacken the five wheel nuts, using the hinged type wheel nut wrench from the vehicle tool kit in the fully extended position. This will provide additional leverage for removal of wheel nuts.
2. Jack up corner of the vehicle—see previous page: jacking the vehicle.
3. Remove the nuts and gently withdraw the wheel over the studs.
4. If available, place a drop of oil or grease on the wheel studs to assist in replacement.
5. Fit spare wheel; tighten the nuts as much as possible with the hinged wheel nut wrench. The extended part will automatically fold to provide normal leverage for refitting. Lower the vehicle to ground and finally tighten the nuts to a torque figure of between 10,0 and 11,7 kgf. m (75 and 85 lbf. ft).

   Note. Do not use foot pressure on extension tubes or wheel studs and nuts may be over-stressed.

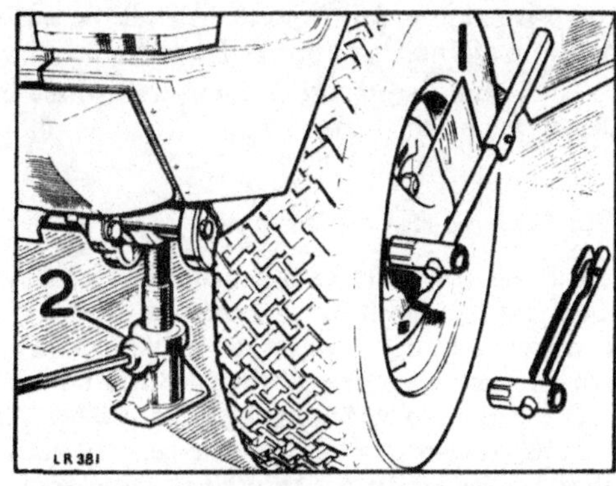

Fig. 74. Wheel changing, front illustrated

Spare wheel removal—**Fig.** 75.

To remove spare wheel from mounting:

1. Unscrew locking lever securing clamping plate and spare wheel, and remove.
2. Remove clamping plate.
3. Remove spare wheel.

Fig. 75. Spare wheel removal

Body care

It is always preferable to clean the bodywork trim with water and sponge, using plenty of water; wherever possible the surface should be freely hosed. After drying with a chamois leather, polish in the usual manner, using any of the good brands of wax car polish.

The use of salt on the roads during frosty weather, sometimes in quite strong concentrations, is now being widely practised. Due to its highly corrosive nature, salt deposited should be washed off as soon as possible by thorough under-washing of the vehicle.

To clean the seats, use a damp cloth with a little mild soap. Do not use detergents on the seats.

Vinyl covered rear quarter panels

Wash the vinyl surface over with warm soapy water (use soap flakes or mild tablet soap). If dirt is ingrained the use of a soft nail brush will help. Rinse off with clean cold water ensuring that all soap is removed. During normal cleaning of the car the vinyl will not be affected by mild detergents such as are used in Automobile Car Washes. Avoid the use of wax polish, creams, solvents or strong detergents. Under no circumstances should White Spirit or Petrol be used to remove oil or grease marks from the vinyl surface.

Air Conditioning

8

WARNING:
The air conditioning system is filled at high pressure with a potentially toxic material. Follow Service instructions when dismantling or applying excessive heat, e.g. steam cleaning, painting, etc. Servicing must only be carried out by a qualified engineer in accordance with instructions in the Repair Operation Manual.

AIR CONDITIONING

The air conditioning system operates in conjunction with the vehicle heating system to provide cooled and dried re-circulated air.

The system delivers cooled or fresh air only to all face-level vents and louvres and fresh or heated air as required to windscreen and footwell outlets.

The installation incorporates temperature and fan speed controls mounted on the facia inboard of the steering column and heating and distribution controls mounted in the central console.

Face-level vents

The two face-level vents can be set to blow cooled or fresh air; each has a knurled knob in the centre which can be rotated to regulate the amount of air delivered. The vent may be adjusted to control the direction of airflow.

Facia-mounted louvres

The five facia-mounted louvres can also be set to blow cooled or fresh air; the vanes may be opened and adjusted to control the direction of airflow.

Air conditioner/temperature control – lower control

For maximum cooling in heavy traffic and for rapid initial cooling, the temperature control should be moved to the 'COLD' position. When the temperature inside the car becomes comfortable, the control should be moved back slightly to prevent the cooling coils from freezing.

Air conditioner/fan control – upper control

The fan control should be adjusted to regulate the volume of air required.

Heater/vent control – left side control

When using the air conditioner, the 'VENT' control should be moved to the 'OFF' position. Refer to 'Fresh air'.

Heater/distribution control – right side control

The distribution control has two positions:
'SCREEN': all heated or fresh air is directed to the windscreen through the demister vents.
'CAR': air, heated or fresh, is directed to the footwells, although a certain amount will continue to flow through the demister vents to the windscreen.

Heater/heat control – upper control

The temperature of the air flowing through the footwell and windscreen vents may be regulated between cold (blue) and hot (red) by moving the control as required.

Effective air conditioning, this control should be maintained in the cold (blue) position.

Heater/air intake control – lower control

For effective air conditioning, this control should be maintained in the 'OFF' position. This prevents the intake of outside air and allows the air conditioning system to process recirculated air.

Full instructions relating to the use of this control are given in 'Heating and Ventilation' Section.

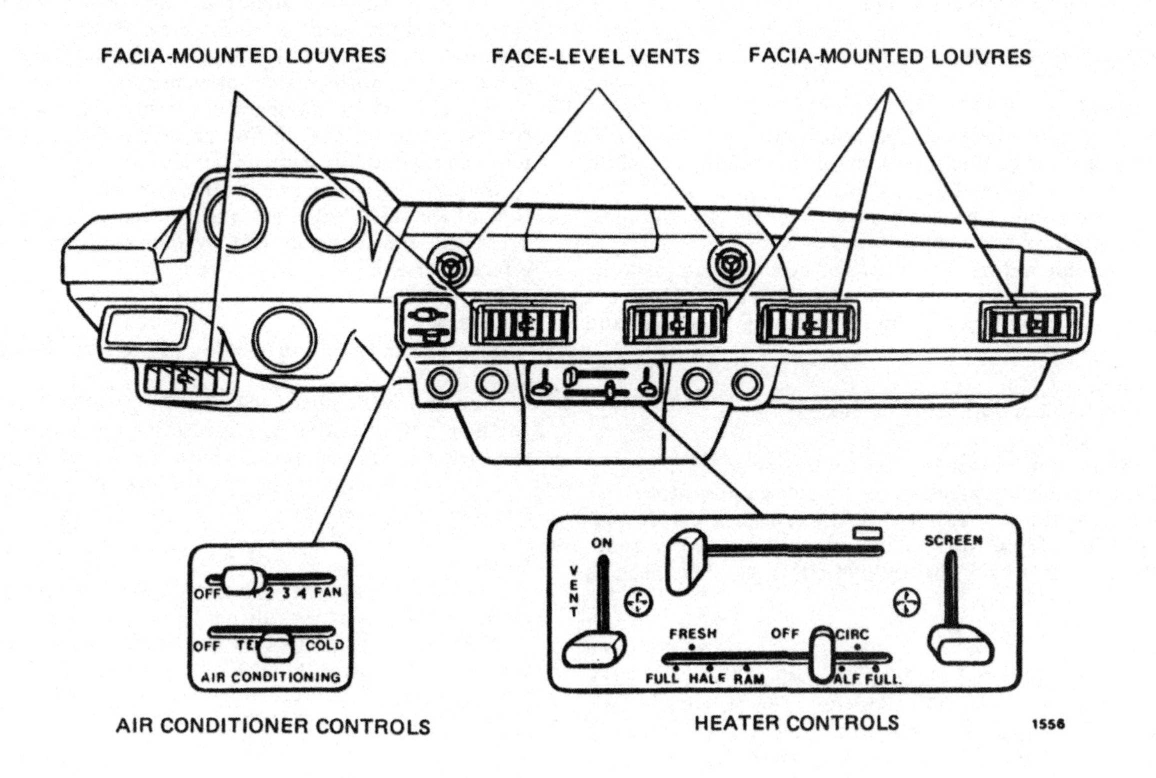

AIR CONDITIONER CONTROLS

HEATER CONTROLS

Air conditioner – turning on

Set the heater controls as follows:
'Vent': 'OFF'.
'Temperature': 'Cold' (BLUE).
'Screen/car': 'CAR'.
'Intake': 'OFF'.

Set the air conditioning controls as follows:

Move the air conditioner fan control to position 1, 2, 3 or 4 to regulate the volume of airflow desired.

Move the air conditioner temperature control to the coldest position for maximum cooling in heavy traffic.

When the temperature inside the car becomes comfortable, move the temperature control back slightly. This will prevent the evaporator cooling coils from becoming too cold and freezing-up.

Rapid cooling

Ensure that all exterior vents are closed.

Open a window.

Move the air conditioner fan control to position 4.

Move the temperature control to the coldest position.

After driving for several minutes, the hot air inside the car will be expelled. Close the window, move the air conditioner temperature control back slightly and adjust the fan speed as desired.

Fresh air

Fresh air may be admitted into the vehicle by moving the heater vent control to the 'ON' position. The air conditioner, operated in conjunction with this function provides an effective means of removing excessive cigarette smoke or stale air whilst maintaining adequate cooling.

Highway driving

During a long trip when the temperature and humidity are extremely high, frost may form on the cooling coils of the evaporator. The unit is equipped with an automatic defrost system which normally will prevent this.

However, if the temperature control is maintained in the coldest position, the defrost system will not operate and the unit cannot supply adequate cold air.

Should this occur, move the temperature control slightly toward the 'OFF' position and the fan control to position '4'. This will allow the defrost system to operate and provide effective cooling.

Demisting

Mist often forms on windows when the humidity is very high.

To remove the mist, move the temperature and fan controls to their low positions. If the interior temperature is too low, use the heater in conjunction with the air conditioner.

It is not necessary to use the system continuously, only when misting persists.

Heating

During cold weather the air conditioner fan can be used to circulate warm air from the heater.

Move the fan control to the desired setting and move the temperature control to 'OFF'.

ROUTINE MAINTENANCE

The design of the system is such that most of the routine servicing is a series of visual checks to be carried out at 10.000 km (6,000 mile) intervals:

Condenser

Using a water hose or air-line, clean the exterior of the condenser matrix.

Check the pipe connections for signs of fluid leakage.

Evaporator

Examine the pipe connections for signs of fluid leakage.

Receiver/drier sight-glass

After running the engine for five minutes, with the air conditioning system in operation, examine the sight glass for signs of bubbles.

Check the pipe connections for signs of fluid leakage.

Compressor

Check the pipe connections for fluid leakage and the hoses for swellings.

Compressor drive-belt

The belt must be tight with not more than 4 to 6 mm (0.19 to 0.25 in) total deflection when checked by hand mid-way between the pulleys on the longest run.

Where belt tension has decreased beyond the limits, a noisy whine or knock will often be evident during operation.

If necessary, adjust as follows:

Slacken the compressor adjuster bolts and pivot bolts.

Adjust the position of the compressor by means of the pivot and slotted fixings to give the correct belt tension.

Tighten all fixings and re-check the belt tension.

It is strongly recommended that any adjustments or rectification procedures should be carried out by your Range-Rover Dealer or an approved automotive air conditioning specialist. Under no circumstances should non-qualified personnel attempt repair or servicing of air conditioning equipment.

See 'Recommended lubricants and fluids' in 'Running Requirements', Section 4 for specified compressor oils and refrigerant.

General Data, Circuit Diagrams, Maintenance Schedules and Index

9

Engine

Type	V8
Bore	88,90 mm (3.500 in.)
Stroke	71,12 mm (2.800 in.)
Number of cylinders	Eight
Cylinder capacity	3528 cc (215 cu in.).
Compression ratio	8.13:1
BHP	156 (116 kw) at 5,000 revs/min.
Maximum torque	28,3 kgm (205 lb ft) at 3,000 revs/min.
BMEP	10,12 kg/cm^2 (144 lb sq in.) at 3,000 revs/min.

BHP, maximum torque and BMEP figures are derived from bench tests and do not allow for installation losses in the vehicle

Firing order	1, 8, 4, 3, 6, 5, 7, 2
Sparking plug	Unipart GSP131 or Champion N12Y 14mm with suppressed leads
Sparking plug point gap	0,80 mm (0.030 in.)
Distributor contact breaker gap	Dwell angle 26°–28°. See also Distributor contact points, page 74
Ignition timing, dynamic at 750 rpm max.	5° ATDC mark on crankshaft pulley—using 91–93 octane fuel—2-star rating in the UK. (Non-emission engines 6° BTDC at 650 rpm max.) 8° ATDC mark on crankshaft pulley—using 85–91 octane fuel. (Non-emission engines 3° BTDC at 650 rpm max.)
Oil pressure	2,11 to 2,81 kg/cm^2 (30 to 40 lb sq in.) at 80 kph (50 mph) in top gear with engine warm (2,400 revs/min.)
Lubrication	Full pressure
Oil filter—internal	Gauze pump intake filter in sump
Oil filter—external	Full-flow

Clutch

Type	Diaphragm spring single dry plate 267 mm (10.5 in.) hydraulic hydrostatic operation
Fluid	Unipart Universal Brake Fluid or other brake fluids having a minimum boiling point of 260°C (500°F) and complying with FMVSS 116 DOT3 or SAE J1703 specification.

Main gearbox

Type .. .. Single helical constant mesh with syncromesh on all forward gears

Transfer box

Type .. .. Two-speed reduction on main gearbox output. Front and rear drive permanently engaged via a lockable differential

Gear ratios

Main gearbox: Top	Direct	
Third	1.505:1	
Second	2.448:1	
First	4.069:1	
Reverse	3.664:1	
Transfer gearbox High	1.113:1	
Low	3.321:1	
Overall ratio (final drive):	In high transfer	In low transfer
Top	3.94:1	11.76:1
Third	5.93:1	17.69:1
Second	9.64:1	28.78:1
First	16.03:1	47.83:1
Reverse	14.43:1	43.07:1

Rear axle

Type .. .. Spiral bevel, fully floating shafts
Ratio 3.54:1

Front axle

Type .. .. Spiral bevel, enclosed universal joints
Angularity of universal joint on full lock .. .. .. 32°
Ratio 3.54:1

Propeller shafts

Type	Open-type, 51 mm (2 in.) diameter, 1310-type universal joints, wide angle variety on front shaft only. Gaiter fitted to sliding coupling of front shaft

Fuel system

Fuel pump	Bendix electrical
Carburetter	Twin Zenith Stromberg type 175 CD-SE identification tag 3854 (non-emission engines 3881)
Needle size	B1EJ (non-emission engines 1EL)
Air cleaner	Replaceable paper element type
Idle speed	700 to 750 revs/min. (non-emission engines 550 to 650 revs/min)
Mixture setting	4.5% maximum at idling speed

Cooling system

Type	Pump, fan and thermostat, pressurised to 1,05 kg/cm² (15 lb sq in.).
Thermostat, wax type	Starts to open at 88°C (190°F) nominal (non-emission engines 82°C (180°F)
Fan belt adjustment	11 to 14 mm (0.437 to 0.562 in.) free movement

Electrical system

Type	Negative earth
Voltage	12 volt
Battery capacity	60 amp hour at 20 hour rate
Ignition system	Coil, 7 volt ballasted, Lucas BA 16C6
Charging circuit	25ACR battery sensed alternator with transistorised current-voltage regulator, output 65 amp
Hazard warning	Switch on dash operates all flashers together
Fuses	35 amp blow rating

Hydraulic dampers	Telescopic double acting non-adjustable 35 mm (1.375 in.) bore
Brakes	
Foot brake	Front: Outboard disc brakes with four pistons Disc diameter 298 mm (11.75 in.) Rear: Outboard disc brakes with two pistons Disc diameter 290 mm (11.42 in.) } Hydraulic servo-assisted self-adjusting
Total pad area	317,34 cm² (49.2 sq in.)
Total swept area	3199,2 cm² (496 sq in.)
Handbrake ('park' brake)	Mechanical 184 mm (7.25 in.) diameter, 76 mm (3 in.) width duo-servo drum brake on rear of transfer box output shaft
Fluid	Unipart Universal Brake Fluid or other brake fluids having a minimum boiling point of 260°C (500°F) and complying with FMVSS 116 DOT3 or SAE J1703 specification.
Steering	
Type	'Burman' recirculating ball, worm and nut
Turns lock to lock	5.55
Front wheel alignment	1,2 to 2,4 mm (0.046 to 0.093 in.) toe-out
Camber angle	0° } Check with vehicle in static unladen condition. That is, vehicle with water, oil and 5 gallons of fuel. Rock the vehicle up and down at the front to allow it to take up a static position
Castor angle	3°
Swivel pin inclination	7°
Steering wheel	Four-spoke: 432 mm (17 in.) diameter.
Tyres	
Sizes	205 x 16 Radial ply (tubed), Michelin M + S, Goodyear 'Wingfoot', or Firestone Town and Country See your Rover Distributor or Dealer for the type of tyre currently recommended

Pressures: Check with tyres cold

Normal on- and off-road use
All speeds and loads

	Front	Rear
kg/cm^2	1,8	2,5
lb sq in.	25	35
bars	1,72	2,4

Off-road 'emergency' soft use
maximum speed of 64 kph
(40 mph)

	Front	Rear
kg/cm^2	1,1	1,8
lb/sq in.	15	25
bars	1,03	1,72

These pressures may be increased for rough off-road usage where the risk of tyre cutting or penetration is more likely. Pressures may also be increased for high speed motoring near the vehicles maximum speed. Any such increase in pressures may be up to an absolute maximum pressure of 2,9 kg/cm^2 (42 lb/sq in.) 2,94 bars.

Normal operating pressures should be restored as soon as reasonable road conditions or hard ground is reached.

After any usage off the road, tyres and wheels should be inspected for damage particularly if high cruising speeds are subsequently to be used.

Replacement bulbs

Headlamps	Lucas No. SP472 60/55/W (Halogen type)	Exterior lamps
Sidelamps	Lucas No. 233, 12v, 4w	Exterior lamps
Stop/tail lamps	Lucas No. 380, 12v, 6/21w	Exterior lamps
Reverse lamps	Lucas No. 382, 12v, 21w	Exterior lamps
Rear fog guard lamps	Lucas No. 382, 12v, 21w	Exterior lamps
Direction indicator lamps	Lucas No. 382, 12v, 21w	Exterior lamps
Side repeater lamps	Lucas No. 989, 12v, 6w	Exterior lamps
Number plate lamps	Lucas No. 233, 12v 4w	Exterior lamps
Instrument panel lamps and warning lamps	Smith No. 4062110974, 12v, 2.2w capless	Interior lamps
Hazard warning switch lamp	Lucas No. 281, 12v, 2w	Interior lamps
Interior roof lamp 'festoon' bulbs	Lucas No. 585, 12v, 10w	Interior lamps
Differential lock warning lamp	Lucas No. 987, 12v, 2.2w	Interior lamps
Clock illumination	Lucas No. 281, 12v, 2w	Interior lamps

Suspension

Front	Coil springs, radius arms and panhard rod. Spring rate: 23,75 kg/cm (133 lb in.)
Rear	Coil springs, radius arms, 'A' frame location arms with 'Boge' hydromat self-energising levelling device. Spring rate: 23,0 kg/cm (130 lb in.)

Dimensions

Overall length	4,47 m (176 in.)	
Overall width	1,78 m (70 in.)	
Overall height	1,78 m (70 in.)	
Wheelbase	2,54 m (100 in.)	
Track: front and rear	1,48 m (58.5 in.)	
Ground clearance; under differential	190 mm (7.5 in.)	
Turning circle	11,28 m (37 ft)	
Loading height	660 mm (26 in.)	
Maximum cargo height	1,04 m (41 in.)	
Rear opening height	1,04 m (41 in.)	
Usable luggage capacity, rear seat folded	1,67 cu m (59 cu ft)	
Usable luggage capacity, rear seat in use	1,24 cu m (43.9 cu ft)	
Vehicle weight: fully laden	2.404 kg (5,300 lb)	
Kerb weight, with water, oil and 22,5 litres (5 gals) of fuel	1.724 kg (3,800 lb)	
Maximum vehicle payload:	680 kg (1,500 lb)	
that is:		
5 persons plus	440 kg (970 lb)	Both on and off the road including auxiliary equipment
2 persons plus	644 kg (1,420 lb)	
Maximum towing weight:	Trailer weight	Trailer plus vehicle weight
Off-road trailer	1.000 kg (2,205 lb)	3.504 kg (7,725 lb)
Road trailer with power brakes	4.000 kg (8,818 lb)	6.504 kg (14,338 lb)
Maximum roof rack load	50 kg (112 lbs)	

Note. It is the Owner's responsibility to ensure that all regulations with regard to towing are complied with. This applies also when towing abroad. All relevant information should be obtained from the appropriate motoring organisation.

Capacities

Component	Litres	Imperial unit	US unit
Engine sump oil	5,1	9 pints	10.5 pints
Extra when refilling after fitting new filter	0,56	1 pint	1.25 pints
Main gearbox oil	2,6	4.5 pints	5.5 pints
Transfer gearbox oil	3,1	5.5 pints	6.5 pints
Rear differential oil	1,7	3 pints	3.5 pints
Front differential oil	1,7	3 pints	3.5 pints
Swivel housing oil (each)	0,26	0.5 pints	0.5 pints
Steering box oil (manual)	0,40	0.75 pints	0.75 pints
Power steering reservoir fluid	1,25	2.2 pints	2.6 pints
Cooling system	11,0	20 pints	24 pints
Fuel tank	81,5	18 gallons	21.5 gallons

Anti-freeze solutions. See page 34 or 35 for anti-freeze recommendations.

Cooling system capacity		Anti-freeze required for 33⅓% solution		Anti-freeze required for 50% solution	
Litres	Pints	Litres	Pints	Litres	Pints
11,3	20	3,7	6.5	5,7	10

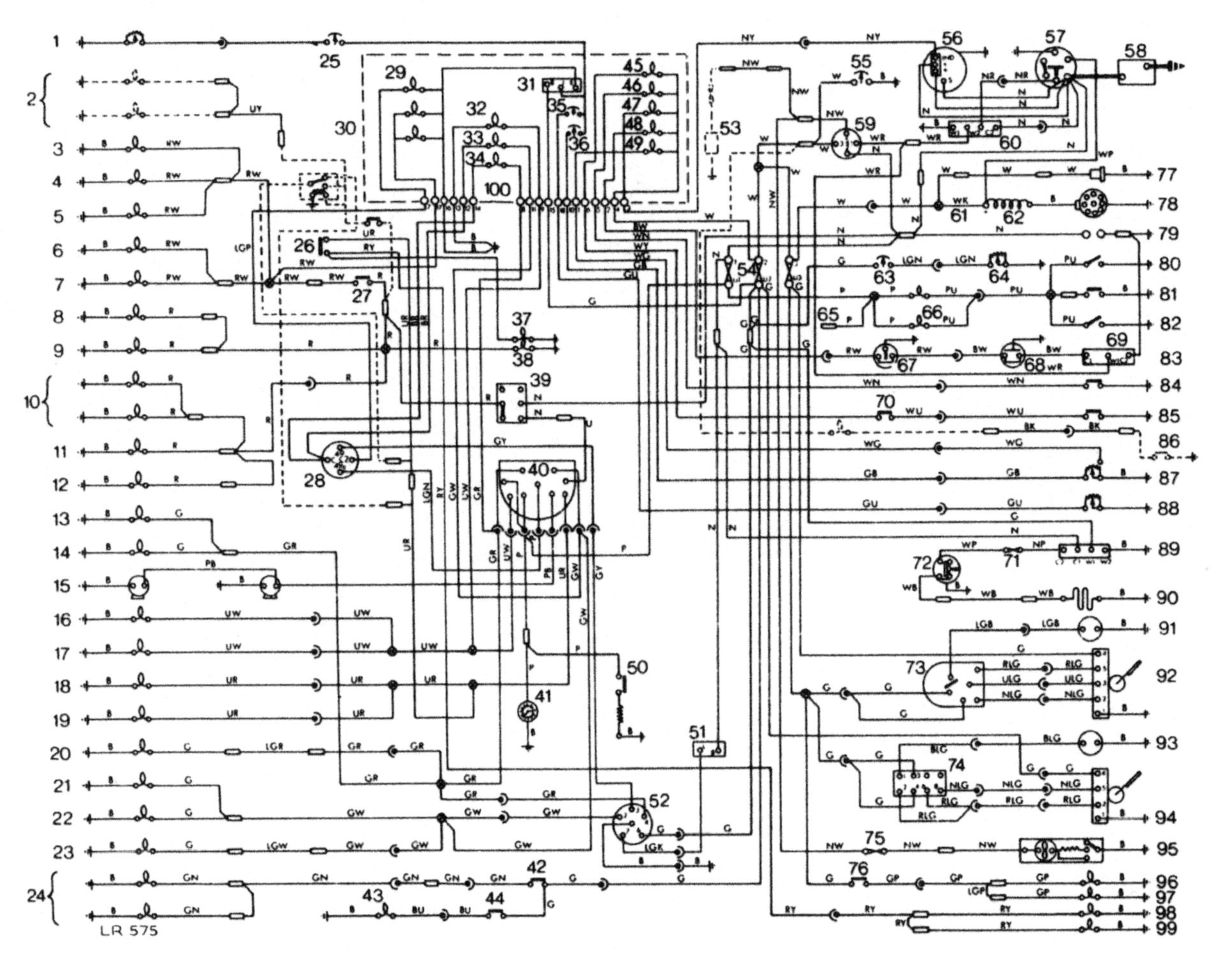

Key to circuit diagram

1 Oil temperature transmitter
2 Pick-up point for front fog lamps
3 Battery voltmeter illumination
4 Oil temperature gauge illumination
5 Cigar lighter illumination
6 Oil pressure gauge illumination
7 Clock illumination
8 Side lamp, LH
9 Side lamp, RH
10 Number plate illumination
11 Tail lamp, LH
12 Tail lamp, RH
13 Side repeater lamp, LH
14 Indicator lamp, front LH
15 Horns
16 Headlamp main beam, RH
17 Headlamp main beam, LH
18 Headlamp dip, RH
19 Headlamp dip, LH
20 Indicator lamp, rear LH
21 Side repeater lamp, RH
22 Indicator lamp, front RH
23 Indicator lamp, rear RH
24 Reverse lamp
25 Oil temperature gauge
26 Switch, rear fog guard lamps
27 Switch, panel illumination
28 Indicator flasher unit
29 Warning light, trailer indicator
30 Panel illumination (two lamps)
31 Voltage stabiliser
32 Warning light, headlamp main beam
33 Warning light, indicator LH
34 Warning light, indicator RH
35 Water temperature gauge
36 Fuel gauge
37 Warning light, rear fog guard lamps
38 Warning light, side lamps
39 Switch, main, vehicle lighting
40 Switch, headlamps, direction indicators and horns
41 Clock
42 Switch, reverse lights
43 Switch, differential lock
44 Warning light, differential lock control
45 Warning light, cold start control
46 Warning light, oil pressure
47 Warning light, ignition
48 Warning light, brake circuit check
49 Warning light, fuel level
50 Cigar lighter
51 Hazard warning flasher unit
52 Switch, hazard warning
53 Pick-up point for radio
54 Fuses
55 Battery voltmeter
56 Alternator
57 Starter motor
58 Battery
59 Switch, ignition
60 Relay, starter motor
61 Ballast resistance wire, coil
62 Ignition coil
63 Oil pressure gauge
64 Oil pressure transmitter
65 Pick-up point for trailer socket.
66 Interior illumination (two lamps)
67 Switch, brake fluid pressure
68 Switch, brake servo vacuum loss
69 Relay, brake check
70 Switch, cold start
71 In-line fuse, heated rear screen
72 Switch with warning light, heated rear screen
73 Switch, front wipers and washer
74 Switch, rear wiper and washer
75 In-line fuse, vehicle heater
76 Switch, stop lamps
77 Fuel pump
78 Distributor
79 Inspection lamp sockets
80 Switch, courtesy light
81 Switch, interior lights (two lamps)
82 Switch, courtesy light
83 Switch, brake circuit check
84 Switch, oil pressure
85 Switch, cold start thermostat
86 'Park' brake (option)
87 Fuel gauge, tank unit
88 Water temperature transmitter
89 Relay, heated rear screen
90 Heated rear screen
91 Windscreen washer motor
92 Windscreen wiper motor, two-speed
93 Rear screen washer motor
94 Rear screen wiper motor, single-speed
95 Heater motor, two-speed
96 Stop lamp, LH
97 Stop lamp, RH
98 Rear fog guard lamp, LH
99 Rear fog guard lamp, RH
100 Printed circuit connector pins

Key to circuit symbols

Snap connectors

Earth connections via fixing bolts

B

Earth connections via cables

Permanent in-line connections

Connections via plug and socket

Key to cable colours

B—Black G—Green K—Pink L—Light N—Brown O—Orange P—Purple R—Red S—Slate U—Blue W—White Y—Yellow

The last letter of a colour code denotes the tracer colour.

ENGINE	Every 5.000 km (3,000 miles) or 3 months	Every 10.000 km (6,000 miles) or 6 months	Every 20.000 km (12,000 miles) or 12 months
Check for oil/fuel/fluid leaks	★	★	★
Check/top up engine oil level	★		
Renew engine oil		★	★
Renew engine oil filter		★	★
Renew engine breather filter			★
Renew carburetter air intake cleaner elements..			★
Check fuel system for leaks, pipes and unions for chafing and corrosion	★	★	★
Check cooling/heater systems for leaks and hoses for security and condition	★	★	★
Check/top up cooling system	★	★	★
Check/adjust operation of all washers and top up reservoirs	★	★	★
Check driving belts, adjust or renew as necessary	★	★	★
Lubricate accelerator control linkage and pedal pivot—check operation		★	★
Check/top up carburetter piston dampers		★	★
Check/adjust carburetter idle settings		★	★
Renew fuel filter element/cartridge		★	★

	Every 5.000 km (3,000 miles) or 3 months	Every 10.000 km (6,000 miles) or 6 months	Every 20.000 km (12,000 miles) or 12 months
Clean/renew engine flame traps			★
Check air intake temperature control system			★
Check crankcase breathing system. Check hoses/pipes and restrictors for blockage, security and condition			★
Clean electric fuel pump element	48,000 miles (80,000 km)		
Check exhaust system for leakage and security	★	★	★
IGNITION			
Clean/adjust spark plugs		★	
Renew spark plugs			★
Clean/adjust distributor contact breaker points		★	
Renew distributor contact breaker points			★
Lubricate distributor		★	★
Check ignition wiring and high tension leads for fraying, chafing and deterioration			★

	Every 5.000 km (3,000 miles) or 3 months	Every 10.000 km (6,000 miles) or 6 months	Every 20.000 km (12,000 miles) or 12 months
Clean distributor cap. Check for cracks and tracking			★
Check security of distributor vacuum unit line and operation of vacuum unit			★
Check/adjust dwell angle and ignition timing using electronic equipment			★
Check coil performance on oscilloscope			★
TRANSMISSION			
Check for oil leaks	★	★	★
Check/top up clutch fluid reservoir	★	★	★
Check clutch pipes for chafing, leaks or corrosion	★	★	★
Check/top up gearbox and transfer box oil levels		★	★
Renew gearbox and transfer box oil	24,000 miles (40,000 km)		
Check front and rear axle case breathers			★
Check/top up front and rear axle oil levels		★	★
Renew front and rear axle oil	24,000 miles (40,000 km)		
Drain flywheel housing if drain plug is fitted for wading	★	★	★
Check tightness of propeller shaft coupling bolts		★	

	Every 5.000 km (3,000 miles) or 3 months	Every 10.000 km (6,000 miles) or 6 months	Every 20.000 km (12,000 miles) or 12 months
Lubricate propeller shaft		★	★
Lubricate propeller shaft sealed sliding joint	24,000 miles (40,000 km)		
STEERING AND SUSPENSION			
Check condition and security of steering unit, joints, relays and gaiters	★	★	★
Check steering rack/gear for oil/fluid leaks	★	★	★
Check shock absorbers for fluid leaks	★	★	★
Check power steering system for leaks, hydraulic pipes and unions for chafing, cracks and corrosion	★	★	★
Check/top up fluid in power steering reservoir *or* manual steering box level	★	★	★
Check/adjust front wheel alignment		★	★
Check security of suspension fixings			★
Check/adjust steering box	★	★	★
Check/top up swivel pin housing oil levels		★	★
Renew swivel pin housing oil	24,000 miles (40,000 km)		
Check suspension self levelling unit for fluid leaks	★	★	★
BRAKES			
Check visually, hydraulic pipes and unions for chafing, leaks and corrosion	★	★	★
Check/top up brake fluid reservoir(s)	★	★	★

	Every 5.000 km (3,000 miles) or 3 months	Every 10.000 km (6,000 miles) or 6 months	Every 20.000 km (12,000 miles) or 12 months
Check footbrake operation (self adjusting)	★	★	★
Check handbrake for security and operation; adjust if necessary	★	★	★
Inspect brake pads for wear, discs for condition	★	★	★
Lubricate handbrake mechanical linkage and cable guides (lever pivot)		★	★
Check brake servo hose(s) for security and condition	★	★	★
Renew hydraulic brake fluid	Every 18 months, 18,000 miles (30,000 km)		
Renew rubber seals in braking system, flexible hoses and servo air filter	Every 36 months, 36,000 miles (60,000 km)		
WHEELS AND TYRES			
Check/adjust tyre pressures including spare wheel	★	★	★
Check tyres for tread depth and visually for external cuts in fabric, exposure of ply or cord structure, lumps or bulges	★	★	★
Check that tyres comply with manufacturers' specification	★	★	★
Check tightness of road wheel fastenings	★	★	★
ELECTRICAL			
Check function of electrical equipment	★	★	★
Check/top up battery electrolyte	★	★	★

	Every 5.000 km (3,000 miles) or 3 months	Every 10.000 km (6,000 miles) or 6 months	Every 20.000 km (12,000 miles) or 12 months
Clean and grease battery connections		★	★
Check headlamp alignment, adjust if necessary	★	★	★
Check, if necessary renew wiper blades	★	★	★
BODY			
Lubricate all locks and hinges (not steering lock)		★	★
Check operation of window controls			★
Check condition and security of seats and seat belts	★	★	★
Check operation of seat belt inertia reel mechanism	★	★	★
Check operation of all door, bonnet and tailgate locks		★	★
Check rear view mirrors for cracks and crazing	★	★	★
Ensure cleanliness of controls, door handles, steering wheel	★	★	★
GENERAL			
Road/roller test and check function of all instrumentation		★	★
Report additional work required	★	★	★

INDEX

Alloy road wheels

Alloy wheels are now available as optional equipment on Range Rovers, or can be obtained through Unipart for vehicles in service in certain markets.

⚠ **WARNING, DO NOT fit alloy wheels in place of steel wheels unless the vehicle is fitted with the latest universal type hubs, see your Land Rover Distributor or Dealer for details. Failure to do this may result in stud failure and LOSS OF ROAD WHEEL.**

Tightening torque for alloy wheel securing nuts: 12,5 to 13,35 Kgf.m (90 to 95 lbf.ft).

LAND ROVER OFFICIAL FACTORY PUBLICATIONS

Land Rover Series 1 Workshop Manual	4291
Land Rover Series 1 1948-1953 Parts Catalogue	4051
Land Rover Series 1 1954-1958 Parts Catalogue	4107
Land Rover Series 1 Instruction Manual	4277
Land Rover Series 1 & II Diesel Instruction Manual	4343
Land Rover Series II & IIA Workshop Manual	AKM8159
Land Rover Series II & Early IIA Bonneted Control Parts Catalogue	605957
Land Rover Series IIA Bonneted Control Parts Catalogue	RTC9840CC
Land Rover Series IIA, III & 109 V8 Optional Equipment Parts Catalogue	RTC9842CE
Land Rover Series IIA/IIB Instruction Manual	LSM64IM
Land Rover Series III Workshop Manual	AKM3648
Land Rover Series III Workshop Manual V8 Supplement (edn. 2)	AKM8022
Land Rover Series III 88, 109 & 109 V8 Parts Catalogue	RTC9841CE
Land Rover Series III Owners Manual 1971-1978	607324B
Land Rover Series III Owners Manual 1979-1985	AKM8155
Land Rover 90/110 & Defender Workshop Manual 1983-1992	SLR621ENWM
(Covering petrol 2.25, 2.5, 3.5 V8 & diesel 2.25, 2.5, 2.5 Turbo Charged)	
Land Rover Defender Workshop Manual 1993-1995	LDAWMEN93
(Covering petrol 2.25, 2.5, 3.5 V8 & diesel 2.25, 2.5, 2.5 Turbo, 200 Tdi)	
Land Rover Defender 300 Tdi Workshop Manual 1996-1998	LRL0097ENGBB
& Overhaul Manuals - 300 Tdi Engine LRL 0070ENG, R380 Manual Gearbox	
LRL 0003ENG, LT230T Transfer Gearbox LRL 0081, Electrical Library	LRL 0126ENG
& Electrical Curcuit Diagrams	LRL 0125ENG
Land Rover Defender Td5 Workshop Manual 1999-2006	LRL0410BB
& Workshop Manual (Relevant Sections) LRL 0097ENG, R380 Manual Gearbox	
LRL 0003ENG & LT230T Transfer Gearbox LRL 0081	

LAND ROVER OFFICIAL FACTORY PUBLICATIONS - continued

Land Rover Defender Electrical Manual Td5 1999-2006 & 300Tdi 2002-2006 Contains - Electrical Circuit Diagrams YVB 101670 & LRL 0452ENG & Electrical Library VDL 100170 & LRL 0389ENG	LRD5EHBB
Land Rover 110 Parts Catalogue 1983-1986	RTC9863CE
Land Rover Defender Parts Catalogue 1987-2006	STC9021CC
Land Rover Defender 90 • 110 Handbook 1983-1990 MY	LSM0054
Land Rover Defender 90 • 110 • 130 Handbook 1991 MY - Feb. 1994	LHAHBEN93
Land Rover Defender 90 • 110 • 130 Handbook Mar. 1994 - 1998 MY	LRL0087ENG/2
Discovery Workshop Manual 1990-1994 (petrol 3.5, 3.9, Mpi & diesel 200 Tdi)	SJR900ENWM
Discovery Workshop Manual 1995-1998 (petrol 2.0 Mpi, 3.9, 4.0 V8i & diesel 300 Tdi) & Overhaul Manuals - 3.9 LRL0164ENG, 4.0 LRL0004 ENG, 300 Tdi LRL0070ENG, R380 Gearbox LRL0003ENG, Transfer Gearbox LT230T LRL0081 & LT230Q LRL0083ENG	LRL0079ENG BB
Discovery Series II Workshop Manual 1999-2003 (petrol 4.0 V8 & diesel 2.5 Td5)	VDR 100090/6
Discovery Parts Catalogue 1989-1998 (2.0 Mpi, 3.5, 3.9 V8 & 200 Tdi & 300 Tdi)	RTC9947CF
Discovery Series II Parts Catalogue 1999-2003 (petrol 4.0 V8 & diesel 2.5 Td5)	STC9049CA
Discovery Owners Handbook 1990-1991 (petrol 3.5 & diesel 200 Tdi)	SJR820ENHB90
Discovery Series II Handbook 1999-2004 MY (petrol 4.0 V8 & Td5 diesel)	LRL0459BB
Freelander Workshop Manual 1998-2000 (K Series 1.8 petrol and L Series 2.0 diesel)	LRL0144
Freelander Workshop Manual 2001-2003 ON (K 1.8L & 2.5L petrol and Td4 2.0 diesel)	LRL0350ENG/4
Land Rover 101 1 Tonne Forward Control Workshop Manual	RTC9120
Land Rover 101 1 Tonne Forward Control Parts Catalogue	608294B
Land Rover 101 1 Tonne Forward Control User Manual	608239
Land Rover Military (Lightweight) Series III Parts Catalogue	
Land Rover Military (Lightweight) Series III User Manual	608180
Land Rover Military 90/110 All Variants (Excluding APV & SAS) User Manual	2320-D-122-201
Range Rover Workshop Manual 1970-1985 (petrol 3.5)	AKM3630
Range Rover Workshop Manual 1986-1989 (petrol 3.5 & diesel 2.4 Turbo VM)	SRR660ENWM & LSM180WS4/2
Range Rover Workshop Manual 1990-1994 (petrol 3.9, 4.2 & diesel 2.5 Turbo VM, 200 Tdi)	LHAWMENA02

Range Rover Workshop Manual 1995-2001 (petrol 4.0, 4.6 & BMW 2.5 diesel) & Overhaul Manuals - 4.0 & 4.6 V8 Engine LRL 0004ENG, R380 Manual Gearbox LRL 0003 & Borg Warner 44-62 Transfer Gearbox LRL 0090ENG	LRL0326ENGBB
Range Rover Workshop Manual 2002-2005 (BMW petrol 4.4 & BMW 3.0 diesel)	LRL0477
Range Rover Electrical Manual 2002-2005 UK Version (petrol 4.4 & 3.0 diesel)	RR02KEMBB
Range Rover Electrical Manual 2002-2005 USA Version (BMW petrol 4.4)	RR02KAMBB
Range Rover Parts Catalogue 1970-1985 (petrol 3.5)	RTC9846CH
Range Rover Parts Catalogue 1986-1991 (petrol 3.5, 3.9 & diesel 2.4 Turbo VM & 2.5 Turbo VM)	RTC9908CB
Range Rover Parts Catalogue 1992-1994 MY & 95 MY Classic (petrol 3.9, 4.2 & diesel 2.5 Turbo VM, 200 Tdi & 300 Tdi)	RTC9961CB
Range Rover Parts Catalogue 1995-2001 MY (petrol 4.0, 4.6 & BMW 2.5 diesel)	RTC9970CE
Range Rover Owners Handbook 1970-1980 (petrol 3.5)	606917
Range Rover Owners Handbook 1981-1982 (petrol 3.5)	AKM8139
Range Rover Owners Handbook 1983-1985 (petrol 3.5)	LSM0001HB
Range Rover Owners Handbook 1986-1987 (petrol 3.5 & diesel 2.4 Turbo VM)	LSM129HB
Engine Overhaul Manuals for Land Rover & Range Rover	
300 Tdi Engine, R380 Manual Gearbox & LT230T Transfer Gearbox Overhaul Manuals	LRL003, 070 & 081
Petrol Engine V8 3.5, 3.9, 4.0, 4.2 & 4.6 Overhaul Manuals	LRL004 & LRL164
Land Rover & Range Rover Technical 'How-To'	
Land Rover & Range Rover Driving Techniques	LR369
Working in the Wild - Manual for Africa	SMR684MI
Winching in Safety - Complete guide to winching for Land Rovers & Range Rovers	SMR699MI

Workshop Manual Owners Edition

Land Rover 2 / 2A / 3 Owners Workshop Manual 1959-1983
Land Rover 90, 110 & Defender Workshop Manual Owners Edition 1983-1995
Land Rover Discovery Workshop Manual Owners Edition 1990-1998

From Land Rover and Range Rover specialists or, direct from Amazon:

www.brooklandsbooks.com

Brooklands Books Ltd., PO Box 904, Amersham,
Bucks, HP6 9JA, England.
www.brooklandsbooks.com

Publication Number: 606917 (Edition 2)

ISBN 9781855201736 Ref: RR20HH 8T8/2748

Printed in Great Britain
by Amazon